THE SPACE SHUTTLE

First published in 2011 by Zenith Press, an imprint of MBI Publishing Company, 400 First Avenue North, Suite 300, Minneapolis, MN 55401 USA

Zenith Press titles are also available at discounts in bulk quantity for industrial or sales-promotional use. For details write to Special Sales Manager at MBI Publishing Company, 400 First Avenue North, Suite 300, Minneapolis, MN 55401 USA.

To find out more about our books, join us online at www.zenithpress.com.

Library of Congress Cataloging-in-Publication Data

Bizony, Piers.
  The space shuttle : celebrating thirty years of NASA's first space plane.
     p. cm.
   ISBN 978-0-7603-3941-1 (hardbound w/ jkt)
   1. Space shuttles—United States—History. 2. Space flights—United States—History. 3. United States. National Aeronautics and Space Administration.   I. Title.
  TL795.5.B594 2011
  629.44'10973—dc23

                                                        2011015001

All photographs are from the archives of NASA unless noted otherwise.

Cover by Simon Larkin

Front cover: Space shuttle *Atlantis*, STS-115, after undocking from the International Space Station on September 17, 2006. Joint operations with the station crew lasted six days.

Back cover:
Crew: *Endeavour*, STS-134; from left, Greg Chamitoff and Andrew Feustel, both mission specialists; Mark Kelly, commander; Greg H. Johnson, pilot; and Michael Fincke and European Space Agency astronaut Roberto Vittori, both mission specialists. April 26, 2011. *NASA/Bill Ingalls*
Launch: *Atlantis*, STS-43. August 1991.
Space: *Endeavour*, STS-61, repairing the Hubble Space Telescope. December 1993.

Printed in China

THE SPACE SHUTTLE

CELEBRATING THIRTY YEARS OF NASA'S FIRST SPACE PLANE

PIERS BIZONY

ZENITH PRESS

CONTENTS

FOREWORD

6

Stage One THE ROUTINE DREAM REINVENTING ACCESS TO SPACE

8

MISSION IMAGES APRIL 1981 – JANUARY 1986

23

Stage Two FALLING FROM GRACE THE REALITIES OF ROCKET FLIGHT

60

Stage Three REBUILDING TRUST AN ERA OF GREAT ACHIEVEMENTS

68

MISSION IMAGES DECEMBER 1988 – NOVEMBER 1994

73

Stage Four UNEXPECTED ALLIES MAKING PEACE WITH RUSSIA

114

MISSION IMAGES FEBRUARY 1995 – OCTOBER 1998

121

Stage Five AN ISLAND IN THE SKY ESTABLISHING THE SPACE STATION

154

MISSION IMAGES DECEMBER 1998 –NOVEMBER 2002

159

Stage Six AT THE CROSSROADS RESHAPING THE SPACE PROGRAM

208

MISSION IMAGES JANUARY 2003 – JULY 2011

215

Stage Seven READYING THE BIRD PROCESSING AN ORBITER

282

Appendix SHUTTLE FLIGHT LOG COMPLETE MISSION SUMMARY

295

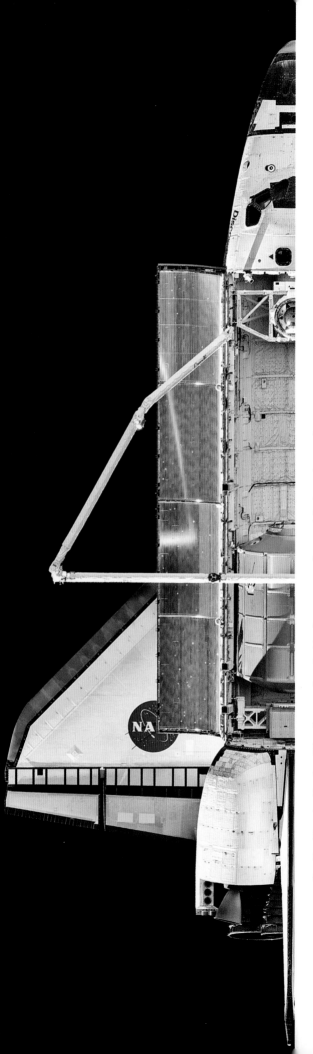

FOREWORD

AFTER THREE DECADES' service and 135 launches, America's Space Shuttle fleet has retired once and for all. The huge launch gantries at the Kennedy Space Center have fallen silent, and the cavernous Vertical Assembly Building must wait for many months, and perhaps some years, before any such grand vehicle is stacked up, once again, within its tall galleries. Thousands of uniquely skilled NASA personnel and contractor staffers have left the scene, all wondering what comes next—not just in their personal working lives, but in the national space program as a whole. The three surviving winged orbiters, *Atlantis*, *Endeavour*, and *Discovery*, are destined for museums, joining test shuttle *Enterprise* in retirement. *Columbia* and *Challenger* were lost in flight, but they will never be forgotten. An era of space capability that we have taken for granted is coming to a close with startling speed.

Since the beginnings of civilization, men and women have ventured to remote and hostile environments, and pushed towards new frontiers, initially by walking, and then by using special means of transport, such as ships, submarines, and airplanes—and now, space vehicles. The glamor of early rocket adventures in the 1960s could not entirely mask the fact that these opening chapters were dominated by political rivalry rather than science and peaceful exploration. Today we have entered into a new era of human space activities: one of international exploration, driven not by Cold War competition but by the quest for

knowledge and the desire to push against the limits of our scientific and technological understanding. NASA's familiar space planes have enabled a permanent human presence aboard the greatest and most ambitious engineering artefact in history: the International Space Station. We have been unforgiving about NASA's occasional failures. A glance at countless science fiction magazine covers from the 1920s onward helps us remember the scope of its achievements across the span of a mere handful of decades. When it comes to the Earth orbital realm, many dreams have been fulfilled. A new artificial world drifts majestically above the sky, from which eerily armored space men and women venture to drift in the cosmic vacuum, adorned by golden helmets. We have an alliance of sixteen nations working in orbital harmony, perhaps presaging the emergence of a global space-faring culture in the not too distant future.

We have created a true world of wonder. We should congratulate everyone concerned, even as we mourn those men and women who have lost their lives in the cause of space exploration. This book is a visual celebration of a flawed yet brilliant spaceship whose kind we may never see again in our lifetimes. It is also a tribute to remarkable human bravery and commitment—on the ground, as well as in space.

Piers Bizony, June 2011

Stage One

THE ROUTINE DREAM

REINVENTING ACCESS TO SPACE

THE ROUTINE DREAM

REINVENTING ACCESS TO SPACE

With the triumph of Apollo's lunar landing completed, America's astronaut program faced an uncertain future. Could the momentum of crewed rocket exploration be sustained into a new era? As political support faded, NASA searched for a more cost-effective and routine vehicle to reach space.

IN THE MARCH 22, 1952, issue of a popular American national magazine called *Collier's*, the German-born rocket pioneer Wernher von Braun described a three-stage winged rocket "approximately the height of a 24-story building." The disposable lower stages lifted a crewed second stage that would return to Earth on wings to be refurbished and flown again. Von Braun certainly popularized the idea of a "space plane," but the aerodynamic math had been done some years before by an Austro-Hungarian engineer, Eugen Sänger. From 1936 and into the World War II years, working under the aegis of the German military authorities, Sänger and his co-worker Irene Bredt conceived a rocket-powered sled on rail tracks that would hurl into the air a winged bomber with its own rocket engines. This would climb into space along a suborbital arc, then repeatedly drop and bounce off the upper atmosphere, shedding energy like a stone skimmed over the surface of a pond. Reaching speeds of 13,000 miles per hour during its descent, the *Silbervogel* ("Silverbird") would span the distance between Germany and the United States in a matter of minutes, then drop its bombs and head for a landing in Japanese stations in the Pacific, or anywhere else that might be prepared to welcome its pilot. The German Air Ministry at that time took a great interest in this and a number of other Sänger schemes, including versions where conventional aircraft "piggybacked" the rocket-powered component high into the air before releasing it for the final rocket-powered push toward the vacuum of orbital space.

After the war, Sänger and Bredt worked for the French government. Sänger's ideas eventually inspired the United States Air Force (USAF) to start work on the X-20 Dyna-Soar, to be launched on a Titan II booster, the missile-derived rocket that fired NASA's two-seater Gemini capsules into space in the mid-1960s. Just like the *Silbervogel*, Dyna-Soar was supposed to travel thousands of miles on a long ballistic arc around the Earth, although its target was Russia, not America. This sinister-looking black craft was the USAF's last bid for the high frontier. In fact, the now-familiar term "aerospace" was coined by public relations consultants seeking to extend the USAF's familiar territory of sky and clouds seamlessly into the vacuum of Earth orbit and beyond, just as the new and competing civilian space agency NASA was taking shape.

Robert McNamara, U.S. Secretary of Defense throughout much of the 1960s, couldn't see the point of military involvement in crewed space activities. Dyna-Soar's supposed role as a bomber seemed irrelevant in an age of intercontinental ballistic missiles, while its other intended use as a spy plane was limited by its relatively short mission profile, especially at a time when orbiting spy satellites were beginning to show their potential. Top-secret Corona satellites (masked behind official press releases about innocent biological research in space aboard "Discoverer" probes) were already taking black-and-white photographs of Russian military facilities. They ejected small re-entry capsules to carry the films back down to Earth. Dyna-Soar never reached beyond an expensive full-scale mockup hidden in a hangar. The project was cancelled in 1962, as the Air Force finally relinquished its grip on human space planning, but the intellectual ghosts of this dark, sleek design lingered in the thoughts of NASA engineers making ready for a post-Apollo world.

M2-F1, known as the Flying Bathtub, with its tow line visible at left, proved that vehicles without wings could make runway landings. This plywood prototype was soon replaced by metal versions capable of high-altitude operations.

Lifting bodies

Even while the first astronauts and cosmonauts amazed the world with their orbital exploits in tiny capsules in the early 1960s, many engineers at NASA dreamed on their own account of creating something half aircraft and half rocket that could return from space and land on a runway. The Mercury, Gemini, and Apollo crew capsules were great at surviving re-entry, but they couldn't fly to controlled landings, and each one had to be recovered from the sea after a final descent and splashdown under parachutes. Having spent their formative years in hard-core aviation, most astronauts were disturbed by this undignified lack of control. Splashdowns were essentially air-sea rescue scenarios. Throughout the 1960s, research was conducted on an alternative crew capsule, shaped like a triangular wedge of cheese with rounded edges, that could fly like a plane, but without wings. It was called a "lifting body." NASA engineer Dale Reed spearheaded the idea from his office at NASA's Flight Research Center at the Edwards Air Force Base in the Mojave Desert of California. As he recalled in a 2002 interview with the author, "The senior managers in NASA didn't believe a

machine without wings could fly, so we built a wooden prototype for $10,000, essentially small change that we didn't have to get approval from head office to spend. Then we towed the vehicle at very low altitude behind a car, a souped-up Pontiac. The first pilot, Milt Thompson, controlled it beautifully."

By the summer of 1963, the wooden lifting body M2-F1, nicknamed the Flying Bathtub, was reaching ever higher in the air, towed on a cable behind a twin-engined Douglas C-47 airplane and then released. It performed amazingly well, and these initial experiments convinced NASA and the Air Force to fund an ever more sophisticated series of metal lifting bodies capable of high-speed, high-altitude flight, sometimes assisted by small rocket motors. They were carried 60,000 feet into the sky, tucked beneath the wing of a converted B-52 bomber. And then they were dropped. Their pilots had only one chance to touch down safely. The program proved beyond a doubt that wingless re-entry capsules could land on a runway so long as they were shaped right. The seed had been sown that would turn into the oak tree of the space shuttle.

THE FIRST SPACE PLANE

In 1954, NASA's precursor organization, the National Advisory Committee on Aeronautics (NACA) designed and oversaw construction of an experimental rocket-propelled plane called the X-15. Its task was to push all known envelopes of power, speed, and altitude in the name of pure scientific and engineering research. It would climb to a new realm sixty miles up in the sky, where the air becomes so thin that conventional wing surfaces lose their purchase and only gas jet thrusters can keep a flying machine under control, until it plunges back again into a region where the word *atmosphere* still has meaning.

In all but name, the thrillingly brutish dart-shaped X-15 was the world's first suborbital crewed spacecraft. It broke many speed and altitude records, flying at over six times the speed of sound and touching down beautifully (almost) every time. It wasn't quite suitable for orbital space flight because its metal skin would have overheated and its thin wings burned off during re-entry from orbital speed of 17,500 miles per hour—the equivalent to nearly twenty-three times the speed of sound. The image at top left shows an X-15 mounted beneath the wing of a specially converted B-52 carrier bomber, just before the little rocket plane is released at high altitude (lower left) for independent flight to the boundary of space, followed by an unpowered glide back down to Earth for a runway landing. The solo pilot had one chance only to touch down safely.

There were many who believed that the X-15 was the natural forerunner for what must surely come next: an orbital space plane. Harrison Storms, head of the Space Division at the X-15's manufacturing company, North American Aviation, pleaded with NASA, "Just let us beef it up a little." Given another few years' development, perhaps an X-15 could have reached orbit, but those years weren't available. The familiar Soviet-American space race took flight on the back of expendable missiles rather than rocket planes. Gagarin, Armstrong, and Aldrin were warrior heroes in a proxy war of thermonuclear display.

An early proposal from the Lockheed company shows the shuttle orbiter mated belly to belly with a reusable carrier stage. Both components are based on lifting body shapes, with the orbiter being almost wingless.

Defining the shuttle

The rocket's red glare soon fades. By June 1969, even as NASA was preparing for its ultimate triumph—the mission of Apollo 11 to the moon—White House aides responsible for liaising with the space agency received a memo from President Nixon's people, warning them of a "concern that NASA and others will use the enthusiasm generated by a success of Apollo 11 to create strong pressures on the President to commit him and the Nation prematurely to a large and continuing space budget." This new president was not willing to build on the Apollo legacy of his old political adversary, John F. Kennedy. "We must devise less costly and less complicated ways of transporting payloads into space," he insisted. His budget request for civilian space activities in 1970 was expected to be around $3.7 billion. The White House warned NASA that "the President is interested in an evaluation of several alternative budget levels, including one in the vicinity of $2.5 to $3 billion."

At first, no one in NASA wanted to believe that the winds had changed. George Mueller, head of the Office of Crewed Space Flight during the Apollo years, was hoping for a piloted mission to Mars as a natural successor to the lunar flights. His boss at that time, NASA's new administrator Thomas Paine, passionately agreed that something impressive should succeed Apollo. The two other most powerful NASA figures at that time, Werhner von Braun at the Marshall Space Flight Center in Huntsville and Robert Gilruth of the crewed Spacecraft Center at Houston, preferred a different objective: an orbiting space station. In 1969, his last year at NASA, Mueller acknowledged that grand missions on the scale of Apollo were a thing of the past and revised his thinking accordingly, even while Paine continued to struggle against Nixon's unambitious expectations for NASA. Mueller began to champion a gradual and more realistic path toward a permanent space infrastructure, centered on a reusable vehicle that would access Earth orbit repeatedly, carrying payloads of people and hardware alike. He and his colleagues called it a "Space shuttle."

Max Faget, a brilliant engineer at the crewed Spacecraft Center, took a leading role in designing the new craft. Together with his longtime collaborator Caldwell Johnson, he had pioneered the blunt-ended conical capsules of the Mercury, Gemini, and Apollo era. Now he proposed two fully reusable stages, a huge delta-winged booster and a smaller component, the shuttle itself, riding piggyback on the booster.

Obviously the skin of the spacecraft has to be heavily shielded against the heat of re-entry, but much more important is the "bow shock" of compressed air that the craft makes as it hurtles through the atmosphere. The streamlined shape of a conventional airplane is designed to reduce atmospheric drag by minimizing bow shocks at normal flying speeds. That's why most jet fighters and airliners look so sleek. But at re-entry speeds of up to 25,000 feet per second, a spacecraft's bow shock needs to be as thick as possible to act as an insulating layer against the furious friction of the surrounding atmosphere. Ordinarily, the bow shocks from wings on a spacecraft would be so narrow that they would just burn away. This had been one of the limiting factors in all space plane concepts to date. Faget turned his shuttle into a capsule-like re-entry object by giving it unusually thick wings with softly rounded leading edges, and by having the entire craft re-enter the atmosphere belly first to create a tremendously fat bow shock, like something right out of the lifting body manual—except that Reed and his colleagues liked the bow shock but weren't impressed by the wings. They'd been trying to eliminate them almost completely. Reed thought they were "parasitic weight on takeoff, because they don't contribute anything, and parasitic in space, because, again, they don't contribute anything, and you can't put fuel or payload inside them. It's better to do away with them."

The problem was that the shuttle envisaged by NASA was supposed to carry payloads back home to Earth as well as hauling them up to orbit. It would have to have large wings to help bear these loads while gliding back, unpowered, through the atmosphere and down to a runway landing. A compromise was reached, and shuttle designs emerged from the half-dozen leading aerospace manufacturers that combined lifting body concepts with wings. In some versions the orbital craft flew on the top of the larger booster, just as Faget and his supporters had recommended. In others, the two components slotted snugly against each other belly to belly for takeoff. Some designs called for a shuttle sitting on top of a cylindrical first stage in a more conventional "in-line" vertical arrangement.

Winning approval

The shuttle certainly looked possible on paper. In the Oval Office and on Capitol Hill, it seemed less so. In May 1970 Nixon responded in writing to NASA's plans. "By no means should we allow our space program to stagnate. But with the entire future and the entire universe before us, we should not try to do everything at once. Our approach to space must continue to be bold, but it must also be balanced." Whatever one's political loyalties, it would not be fair to suggest that Nixon was entirely unique in his distaste for more rocket adventures now that Apollo had done its job. Times had changed, and so had national priorities. In that same month, Walter Mondale, the Democrat senator for Minnesota, spoke passionately in the Senate against NASA's budget request for the first year's work on developing the shuttle. "I believe it would be unconscionable to embark on a project of such cost when many of our citizens are malnourished, when our rivers and lakes are polluted, and when our cities and rural areas are dying. What are our values? What do we think is more important?"

NASA administrator Paine hoped that Nixon's administration would allow the savings from three cancelled Apollo lunar landing missions (18, 19, and 20) to be allocated for the shuttle's initial development. Incidentally, those cancellations are often laid at the feet of the Nixon administration, but there were many senior figures within NASA who were anxious to bring Apollo to a close, fearful of the enormous risks involved in each one of those lunar missions. That said, Nixon was not a NASA fan. Paine could make no headway and resigned from the chieftainship of America's space program after just two years. In 1986 he explained, "I finally left because I didn't think I could deliver the kind of relationship with the President that the head of NASA really ought to deliver." A new administrator, James Fletcher, was appointed by Nixon to redirect NASA away from what the White House termed "rapid razzle-dazzle growth and glamor" and toward "organizational maturity and more stable operations for the long-term." In other words, Fletcher's task was to stomp on the big dreams of the rocket cadets and bring space activities under the tight disciplines of fiscal control.

Fletcher complied as best he could. During his tenure, the tough decision was made to dispense with the shuttle's piloted and fully reusable winged booster stage in favor of simpler and only partially reusable solid rockets, and to build the shuttle orbiter around a conventional aluminum airframe instead of titanium: an

▶
Test pilots jape around an HL-10 lifting body. This flight test program, conducted throughout the late 1960s, delivered useful data that has not yet been fully exploited. Lifting bodies make good re-entry vehicles, but this concept was overshadowed by simpler rocket designs.

THE RISKS OF WINGLESS FLIGHT

As with all high-speed aerospace experiments, there were unknown hazards ahead, both for the X-15 rocket plane (which claimed one fatality) and the lifting bodies. Until his death in May 2006 at the age of seventy-two, former NASA research pilot Bruce Peterson always looked very dashing in retirement, with his black eye patch, brown flying jacket, chinos, and boots. In 2001 the author met him on the flat dry lake bed that makes up the flight line at Edwards Air Force Base.

Laconically, Peterson related his crash in a fully equipped metal version of Reed's lifting body, dropped from 40,000 feet: "It was September 27, 1968. What happened was, the standby rescue helicopter pilot didn't attend the briefing on the morning of my flight, and he got in the way of my ship. I lost control because this was the first flight of an untested design, not quite the same shape as the wooden prototype that had done so well previously, and this thing didn't fly too well." The craft got into a

bad roll motion as it plunged toward the lake bed. "I was late putting the landing gear down by about one and a half seconds," Peterson admitted. "I hit the floor and rolled over several times, and the ship ended upside-down. And that was the end of the ball game."

Archive footage shows this, and later, the rescuers standing around with their hands in their pockets, gloomily surveying the scene of the disaster (as in the photo below) in the certain knowledge that Peterson must be dead. Then, according to Peterson, the rescuers eventually realized he was still alive and cut his battered body out of the wreckage. Footage of the spectacular crash, complete with errant helicopter, featured in the opening credits of the popular 1970s TV adventure show, *The Six Million Dollar Man*. Peterson's story reminds us of the bravery and raw determination behind the making of the shuttle. Lifting body research was essential to its genesis..

even lighter and far more heat-resistant metal that would have been an excellent material had it not been for the additional costs of machining and shaping this unusually tough stuff into the needed components. This was a crucial turning point in the shuttle's genesis. A fully reusable system that would have cost a hefty $10 billion in 1970s dollars to develop but might have been

relatively inexpensive to fly and refurbish between flights became, instead, a partially reusable system that was projected to cost less than $6 billion, yet would turn out to be expensive to service and fly.

Fletcher introduced what seemed at the time a bold and practical initiative: a plan to make the shuttle available as a "workhorse" for private industry as well as

for space scientists. NASA would offset a proportion of the ship's running costs by launching payloads for other federal agencies and commercial customers. Fletcher reinforced the fragmented political support for a continued crewed space program by promising cheaper access to space for communications satellites, university experiments, and military payloads. This was certainly an appealing prospect for Nixon. It seemed a way of turning NASA into a genuine national resource instead of just an expensive club for space cadets. Fletcher called upon cost-benefit analyses from a respected firm of consultants to justify the shuttle as the next major national space program. These particular consultants, based in the prestigious university town of Princeton, New Jersey, were (and still are) justifiably renowned in the fields of business, economics, and public affairs. It might be argued that they were less sure-footed in their understanding of rocket vehicles and the people who build and fly them. They suggested to Fletcher that thirty-nine flights per year, with up to five hundred flights conducted between 1978 and 1990, transporting humans and cargo alike in the same vehicles, would make the shuttle competitive with other launch systems, both crewed and uncrewed.

No one with any actual experience of the space business believed in these figures. Fletcher was simply doing what had to be done in order to win political backing for NASA's new ship. Despite the consultants' beguiling business-speak and the drastic reductions in the scale and ambition of the beautiful two-stage shuttle that Faget and Mueller had imagined, still the Nixon administration balked at recommending the project to Congress. Caspar ("Cap") Weinberger, at that time a rising power at the Office of Management and Budget, intervened to persuade Nixon that NASA's requests couldn't be shelved indefinitely. Some kind of commitment had to be made for the sake of national prestige. On August 12, 1971, Weinberger sent Nixon a single short memo that probably saved the shuttle from a lingering death of a thousand cuts. He argued that there was a danger of giving the impression that "our best years [in space] are behind us, and that we are turning inward." He further stated, "NASA's proposals have some merit. The real reason for sharp reductions in NASA's budget is that we cut it because it is cuttable, and not necessarily because it is doing a bad job."

In an era of economic downturn after the boom years of the 1950s and '60s, Nixon knew that he could not count on popular support for a new rocket program now that Apollo's task was done, but he certainly did not wish to be remembered as the president who killed crewed space exploration altogether. So he scrawled a brief note in the margins of Weinberger's memo. "I agree with Cap." In January 1972, after a welcome private meeting at Nixon's private home in San Clemente, California, Fletcher won the go-ahead for NASA to build the new craft. Nixon stated proudly, "I have decided today that the United States should proceed at once with the development of an entirely new type of space

If more money had been available for the shuttle system's development, a fully reusable piloted first stage would have carried the orbiter to the edge of space, as in this artwork inspired by the work of Eugen Sänger.

transportation system designed to help transform the space frontier of the 1970s into familiar territory, easily accessible for human endeavour in the 1980s and '90s." The NASA delegation had a hard time wresting the cute little scale model of the shuttle out of Nixon's hands. It was a good sign, Fletcher thought. And that's what NASA called it: a "Space Transportation System" (STS).

Secret shuttle services

Other, covert influences were a factor in the decision-making processes. During 1969 the National Reconnaissance Office (NRO)—an organization so secret not even its name would have been confirmed by any government official at that time—requested that the shuttle's payload bay should be much longer and wider than NASA originally had in mind. The NRO built and operated large, expensive reconnaissance satellites, funded by budgets so "black" even Congress was not allowed to check the expenditures, lest their scale give away sensitive information about the actual nature of the satellites. Anyway, NASA made the changes, and the Air Force eventually came on board too, similarly favoring a large payload bay. By 1979 it had even begun construction of a second shuttle launch complex at Vandenberg Air Force Base in northern California, suited specifically for reaching polar orbits. (It was never completed.) NASA entered many long and tortuous agreements with military and intelligence agencies whose principal contribution was to lend their tacit approval, or at least, not stand in the way, when NASA asked the White House and Congress for the necessary financial backing.

All these details might seem little more than long-forgotten bureaucratic minutiae, unimportant some forty years after the fact, but those early satellite reconnaissance requirements, based largely on the military and intelligence communities' vague instincts

about their future needs rather than on hard facts, determined the size of the shuttle's payload bay—and from those decisions, everything else flowed, including the length and mass of the Hubble Space Telescope and even the diameter of the International Space Station's modules. Government space vehicles are engineered as much by politics as they are by rocket experts.

The "most complex machine"

Throughout the 1970s there were several changes of administration in the White House. The Watergate scandal ended Nixon's occupancy in 1974; Gerald Ford's caretaker presidency gave way to that of Democrat Jimmy Carter, who was subsequently undone by the Iranian hostage crisis. All these political dramas overshadowed NASA's fortunes. Meanwhile, the shuttle proved to be just as complex to design and build as Apollo had been. After due competitive process and evaluation of rival costings and concepts, the task of constructing the winged orbiter was awarded to the Rockwell company, the new corporate guise of North American Aviation, former builders of the Apollo capsule. The large external fuel tank went to Martin Marietta, and the solid rocket boosters on the sides of that tank became the province of Morton Thiokol, a company with a proven record of excellence in solid rocket technology.

Some veterans of the Apollo era were concerned about the use of solid rockets on a human-carrying vehicle. It was unprecedented. Once lit, solid motors

THE FAITHFUL SHEPHERD

The Boeing 747-100 Shuttle Carrier Aircraft (SCA) was a hero in its own right. It was originally owned by American Airlines, whose livery was still partially in evidence on the plane's flanks for the first few years of its operation until the space agency finally imposed its own paint scheme in 1983. It was purchased in 1974, at which point NASA technicians stripped out all the passenger seats and cabin fittings, beefed up the engines, strengthened the fuselage roof, and adapted this reliable airliner for use as a cross-country ferry for shuttle orbiters that touched down at the Edwards Air Force Base landing strip in California at the end of their missions. The combined 747-orbiter hardware made an impressive sight heading back to the Kennedy Space Center in Florida. In 1988 another plane of closely equivalent specification, a 747-100SR, was purchased from Japan Airlines and similarly adapted.

cannot be shut off again. Neither are they amenable to throttle control. It seemed to many commentators that this would limit a shuttle crew's options for escape in the event of a launch problem. Even so, there was no doubting Thiokol's expertise. In retrospect, the solid rocket failure that led to the *Challenger* disaster of 1986 lay more in the misuse of that hardware in unsuitable launch conditions, rather than in any fundamental design flaw in the hardware itself.

Dan Dumbacher was on the team developing the shuttle's trio of reusable hydrogen-fueled main engines. "It was not an easy problem to make engines that could last many flights instead of just surviving a few minutes and then getting thrown away forever into the Atlantic. First, you have super-cold liquid hydrogen and oxygen at minus 400 degrees Fahrenheit flooding in. Then, fractions of a second later, you're combusting these fuels at plus 5,000 degrees. You're lucky if you can keep the engine's outer structure at a modest 1,000 degrees to stop it melting. And you have to force-feed fuel at enormous pressure. The turbine fuel pumps have blades that spin at 35,000 revolutions per minute. That's twice the speed of a jet engine."

John Young, the astronaut destined to command the first orbital shuttle mission, knew from the start that

this was not an easy machine to build. In 2006 he told a NASA interviewer that he recalled "Dr. Gilruth telling me it's going to be as reliable as a DC-8." It was common for NASA to sell the dream of a space vehicle being as trustworthy as that old Douglas, but for Young, the problem was that "every time we went out to Rocketdyne or somewhere to see what was happening, engines were blowing up. So I wasn't sure it was going to be as reliable as a DC-8."

The flight characteristics of the combined vehicle during ascent from the ground were also unusual. Instead of one Saturn-style rocket stack, there would be a winged orbiter, a fat fuel tank, and two side-slung solid rocket boosters all lashed together in parallel, with each

of the four components exhibiting distinct and sometimes clashing aerodynamics. The last and seemingly most intractable problem was the orbiter's heat shielding. It had to be lightweight and reusable, yet all-encompassing. The now familiar ceramic tiling system gradually evolved as a delicate and labor intensive solution, capable of leaving dangerous gaps in the thermal protection after the slightest mechanical disturbance. Overall, NASA and its contractors took on a heavy load of new things to do. No one had made reusable rocket engines before—nor a winged orbital vehicle, nor a machine with a payload bay that could open and close. On and on went the new and unprecedented requirements. Up and up the budget climbed. But they did it—and at last, a shuttle took to the air for a taste of reality rather than mere hope.

On the shoulders of giants

Back in mid-1968, when NASA was facing serious tragedies, delays, and cost overruns in its Apollo program, the first successful Apollo-Saturn missions dispelled any doubts about that new technology and restored NASA's prestige in the wake of the January 1967 launchpad fire that had killed the Apollo 1 crew, Gus Grissom, Roger Chaffee, and Ed White. Likewise, the Shuttle Approach and Landing Test Program (ALTP) in 1977, during which an almost fully equipped orbiter named *Enterprise* (its rocket engines were dummies) was shrugged off the back of a converted Shuttle Carrier Aircraft (SCA) to make its own independent landing, burned off the clouds of despondency that had hung over the shuttle during its long and tortuous development. The ALTP was conducted from February through November 1977 at the Dryden Flight Research Center in California. Two NASA astronaut crews, Fred Haise with Gordon Fullerton and Joe Engle with Dick Truly, took turns flying the 150,000-pound spacecraft to

unpowered landings. Five captive flights, with *Enterprise* mounted, uncrewed, atop the SCA tested the basic structural integrity and performance of the mated vehicles. Three crewed captive flights followed, with astronaut crews cautiously testing some of the control surfaces while *Enterprise* was still safely mated to the mother ship. On August 12, 1977, with Haise and Fullerton on board, *Enterprise* separated from the SCA in mid-air and flew on its own for the first time, making a smooth runway landing on the dry lake bed at Edwards Air Force Base. Four more such flights concluded the ALTP series, with the final touchdown taking place on a concrete runway: a specific target with markedly less room for error than the broad open spaces of the natural lake bed.

In March 1978, the *Enterprise* was ferried atop the SCA to NASA's Marshall Space Flight Center, where it was mated with an external tank and solid rocket boosters for a series of vertical ground vibration tests. A similar test was conducted at the Kennedy Space Center a few months later. But this orbiter was never destined to reach space—though it did visit Paris for the International Air Show in 1983.

THE SPACE SHUTTLE FLEET

Enterprise—named for the world's most-loved fictional starship, after a successful campaigning effort from thousands of *Star Trek* fans.

Columbia—named in honor of the first American sailing vessel to circumnavigate the globe in 1836.

Challenger—named after a navy exploration vessel of the 1870s that charted vast regions of the Atlantic and Pacific oceans.

Discovery—named after the vessel in which Henry Hudson and his crew discovered the Northwest Passage between the Atlantic and Pacific oceans.

Atlantis—named for a two-masted ketch operated by the Woods Hole Oceanographic Institute, MA, from 1930 to 1966, and which traveled more than half a million miles in its long career.

Endeavour—named after the British Navy vessel HMS (His Majesty's Ship) *Endeavour*, the ship which took Captain James Cook on his first voyage of discovery (1768–1771).

Columbia was the first shuttle to be delivered to NASA's Kennedy Space Center in March 1979, fully equipped for orbital flight. *Challenger* was delivered in July 1982 and *Discovery* in November 1983. *Atlantis* was rolled out in April 1985, while *Endeavour*, built as a replacement following the 1986 *Challenger* accident, was delivered in May 1991. The first orbiter, named *Enterprise*, never flew in space but was used for approach and landing tests at the Dryden Flight Research Center and several launch pad studies in the late 1970s. Assembly work on *Enterprise*, officially known as Orbiter Vehicle-101 (OV-101), began in June 1974. It rolled out of the fabrication hangar at Palmdale, California, on September 17, 1976. This was the first time that the press and public saw NASA's new vehicle.

Inaugurating a new era

When, at long last, commander John Young and pilot Robert Crippen strapped themselves into shuttle *Columbia* for its inaugural flight into space on April 10, 1981, it had been six years since America last sent astronauts into orbit. Inside the cabin, five computers cross-checked each other's results. For safety, at least four of them had to match data before *Columbia* could fly, but twenty minutes before the scheduled liftoff, they were not in accord. Young and Crippen had to climb out of the flight deck for that day. A million people camped out on nearby beaches were told they would have to pitch their tents for another couple of days before they could see the launch.

At last, on April 12, *Columbia* lifted off the pad and ascended flawlessly. NASA was back in business and the public was ecstatic. Once in orbit, Young and Crippen experienced the benefits of *Columbia*'s large cabin, which was much roomier than the old Apollo capsules' had been. There was even a second deck beneath the cockpit with an airlock, sleeping berths, and a private washroom cubicle. Behind the cabin was a payload bay sixty feet long and fifteen feet wide. One day soon it would carry pressurized space laboratories, satellites, or planetary probes. For now, it was empty.

For this first flight, Young and Crippen were strapped into ejection seats. Once the shuttle had proven reliable, these would be replaced with seven crew couches (four on the main deck and three down below). But was the ship safe yet? When the two men looked out of the rear-facing cargo bay windows, they were in for a shock. Sixteen heat shield tiles had fallen off during launch and as it transpired later, at least a hundred more were slightly damaged. According to Crippen, "Our own engineers said that a few of the tiles on our underside had probably fallen off. If enough had, they said, then the plasma heat from our re-entry would burn right through us." NASA asked the U.S. Air Force to point one of its secret spy satellites at the shuttle while it presented its vulnerable underbelly for inspection. The photos were never released, although it was made clear to NASA that none of the most crucial tiles was missing.

After two days in space, *Columbia* headed back to Earth. Crippen described the moment: "We hit the atmosphere at around twenty-four times the speed of sound. After a while, we could see a pink glow outside our windows, but other than that, there absolutely was no sensation of heat." *Columbia* then used its large wings to glide, unpowered, to a perfect landing at Edwards Air Force Base in the Californian desert, a famous testing ground for new aerospace vehicles. Unlike with a conventional plane, there was no way to turn around during the final approach. *Columbia* had to touch down correctly on the first try, guided by its computers and the occasional nudge from a human hand on the controls. Crippen was thrilled. "The shuttle was a joy to fly. It did exactly what you told it to, even in unstable regions of air." It looked like a typically smooth NASA adventure, and unlike many space missions today, *Columbia*'s debut made headlines around the world. Over the next five years, sister ships *Challenger* and *Atlantis* joined *Columbia* on active service. All three shuttles achieved dazzling successes during what appeared to be an increasingly routine schedule of orbital missions.

NASA celebrated its flagship spacecraft as "the most complex machine in history." In those first, innocent five years of orbital triumphs, few people outside the space industry realized that this fabulous complexity was not a virtue but a hazard. Judging by surface appearances, the shuttle was a science fiction dream come true, an awesome winged spaceship flying time and again into orbit and coming home to land like a plane. Nothing was more certain to turn dreams into nightmares than America's confidence that a machine powered by five infernos of hydrogen, oxygen, aluminum powder, and ammonium perchlorate could be considered so much tamed that its launches were barely worth remarking upon any more.

Enterprise **was named for the world's most famous TV starship. Seen here during a rollout ceremony in September 1976 are, from left to right, NASA Administrator James D. Fletcher, DeForest Kelley (Dr. McCoy), George Takei (Mr. Sulu), James Doohan (Scotty), Nichelle Nichols (Lt. Uhura), Leonard Nimoy (Mr. Spock),** Star Trek **producer Gene Rodenberry, an unidentified man, and Walter Koenig (Ensign Checkov).**

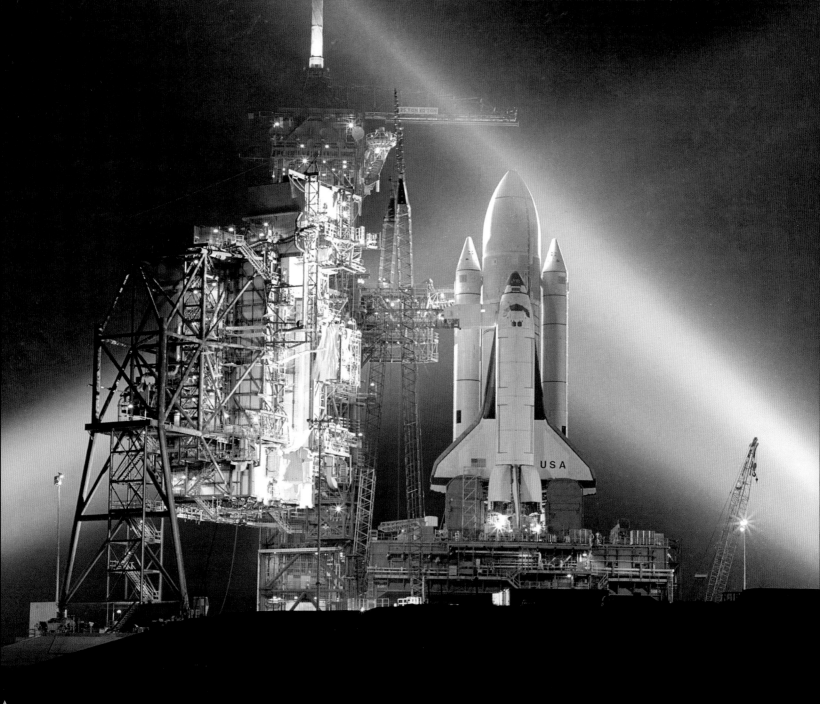

Space Shuttle Columbia *on the pad, bathed in artifical light,
as preparations for the system's inaugural flight to orbit get
under way during the first days of April 1981.*

▶

Lift-off for Columbia, *on a mission designated STS-1.
A new era of space transportation begins in an optimistic
blast of flame and thunder.*

STS-1 *Columbia*

Launched	April 12, 1981
Landed	April 14, 1981
Commander	John W. Young
Pilot	Robert L. Crippen

Milestones

This debut crewed mission for the space shuttle
system was also was the opening test of the
launch stack as a whole: an unusual step for a
human-carrying rocket vehicle that had not first
been flown as an uncrewed prototype. In
acknowledgement of this mission's status as an
experimental flight, escape hatches were fitted

directly overhead Young and Crippen's crew
stations, and their couches were replaced with
ejection seats. *Columbia* lost sixteen insulation
tiles during launch, but its re-entry was not
endangered. America's first human space mission
since 1975 was greeted with jubilation.

Young occupies the commander's left-side seat, with Crippen in the right-hand station. This photo was taken during a training simulation prior to the actual STS-1 mission, but it gives a good impression of their enviroment once in space.

Crippen floats, weightless, in Columbia's mid-deck, making full use of the generous space within the orbiter's pressurized crew compartment. Previous capsules (Mercury, Gemini, and Apollo) were very cramped by comparison.

Surely the ultimate astronaut, John Young, seen here during the STS-1 mission, is the only person in history to have flown three different kinds of spacecraft: Gemini, Apollo, and the shuttle. (He also walked on the Moon.)

▲
Orbiter Columbia *glides to an unpowered runway landing at Edwards Air Force Base at the conlcusion of the STS-1 mission. Apart from a few loose tiles, the flight was perfect.*

◄
Gaps in Columbia's *thermal protection system, created by launch vibrations loosening fragile tiles, are clearly visible on the starboard Orbital Maneuvering System pod.*

Relieved mission controllers at the Johnson Space Center
near Houston monitor the successful touchdown of STS-1.

STS-2 *Columbia*

Launched	November 12, 1981
Landed	November 14, 1981
Commander	Joe H. Engle
Pilot	Richard H. Truly

Milestones

A dangerous spillage of hypergolic nitrogen tetroxide while pad technicians were fueling *Columbia*'s maneuvering thrusters (the Reaction Control System) was just one among several frustrating glitches forcing a three-week launch delay. Payloads included an air pollution detector and an imaging radar experiment, but the shuttle was still in test flight phase. Its main payload was the Development Flight Instrumentation (DFI) pallet, a suite of instruments monitoring every aspect of the vehicle's performance. The planned five-day mission was cut by three days after the sudden failure of one of the three fuel cells that provided *Columbia* and its crew with electricity and drinking water.

◀
STS-1 Columbia *on the runway at Edwards Air Force Base (AFB) after a smooth re-entry. The maintenance trucks are "safing" the orbiter by pumping out the last traces of toxic hypergolic fuels from its tanks.*

▲
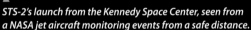
STS-2's launch from the Kennedy Space Center, seen from a NASA jet aircraft monitoring events from a safe distance.

◀
An excellent view of Columbia's *heat-resistant belly as it comes in to land at the end of mission STS-2. The black tiles are the most crucial defense against the heat of re-entry.*

STS-3's Plasma Diagnostics Package (PDP) is grappled by Columbia's Canadian-built robot arm: one of the most important and successful elements of the shuttle system.

STS-3 *Columbia*

Launched	March 22, 1982
Landed	March 30, 1982
Commander	Jack R. Lousma
Pilot	C. Gordon Fullerton

Milestones

The Canadian-built Remote Manipulator System stretched its slender boom arms beyond the confines of the payload bay for the first time in space. The main payload consisted of pallet-mounted experiments for NASA's Office of Space Science. The ejection seats and flight monitoring instruments were still in place for what was essentially the third test flight for this new vehicle.

▶

A huge crowd gathers to watch STS-4's homecoming at Edwards AFB in California. Public interest in the shuttle was at its peak in those early years.

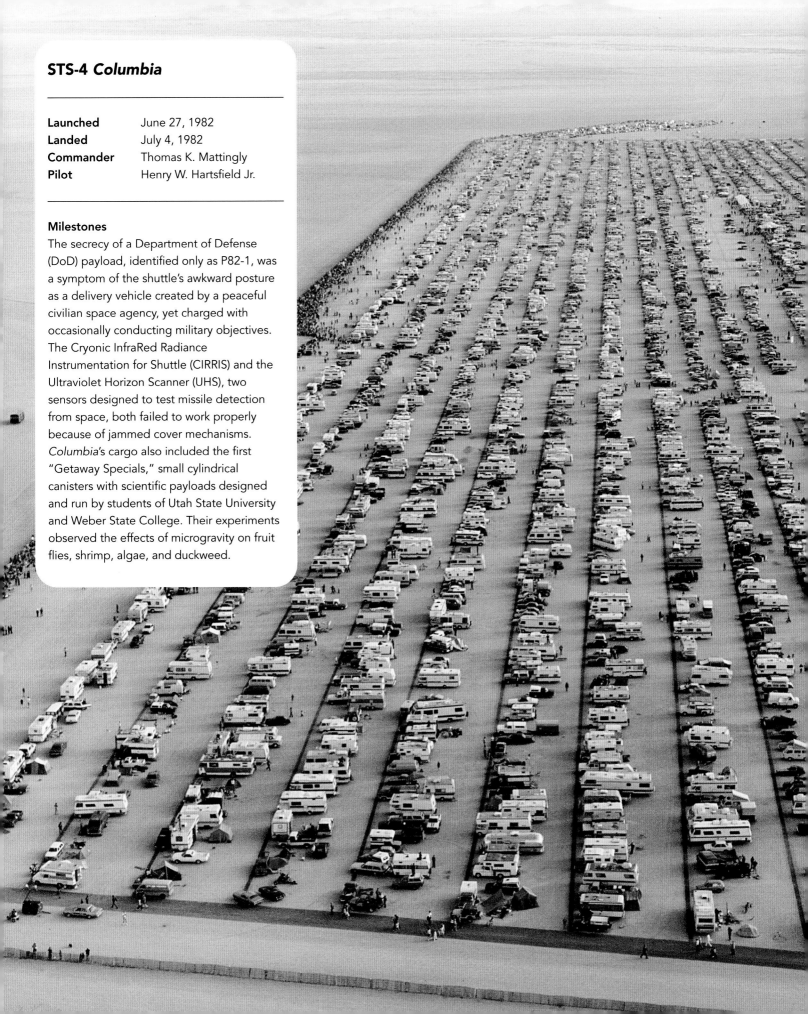

STS-4 *Columbia*

Launched	June 27, 1982
Landed	July 4, 1982
Commander	Thomas K. Mattingly
Pilot	Henry W. Hartsfield Jr.

Milestones

The secrecy of a Department of Defense (DoD) payload, identified only as P82-1, was a symptom of the shuttle's awkward posture as a delivery vehicle created by a peaceful civilian space agency, yet charged with occasionally conducting military objectives. The Cryonic InfraRed Radiance Instrumentation for Shuttle (CIRRIS) and the Ultraviolet Horizon Scanner (UHS), two sensors designed to test missile detection from space, both failed to work properly because of jammed cover mechanisms. *Columbia*'s cargo also included the first "Getaway Specials," small cylindrical canisters with scientific payloads designed and run by students of Utah State University and Weber State College. Their experiments observed the effects of microgravity on fruit flies, shrimp, algae, and duckweed.

STS-5 Columbia

Launched	November 11, 1982
Landed	November 16, 1982
Commander	Vance D. Brand
Pilot	Robert F. Overmyer

Mission specialists
Joseph P. Allen, William B. Lenoir

Milestones

Brand was a veteran of the 1975 Apollo-Soyuz docking mission, the first joint space program between Russia and the United States. The ejection seats were removed and two more seats installed on the flight deck for this first fully operational shuttle mission, which deployed two commercial communications satellites, ANIK C-3 for TELESAT Canada and SitS-C for Satellite Business Systems. The first scheduled space walk of the shuttle era was canceled because of a malfunction in a space suit.

The ANIK C-3 satellite is deployed into space atop its small rocket motor package.

Orbiter Columbia *reflected in a pool of rainwater after touchdown at the end of STS-5.*

STS-6 *Challenger*

Launched	April 4, 1983
Landed	April 9, 1983
Commander	Paul J. Weitz
Pilot	Karol J. Bobko

Mission specialists
F. Story Musgrave, Donald H. Peterson

Milestones
The payload was the first Tracking and Data Relay Satellite-1 (TDRS-1), part of a communications infrastructure to support future NASA missions. The first successful space walk—or Extra-Vehicular Activity (EVA)—of the shuttle program, a four-hour session by Peterson and Musgrave (seen here in a spectacular photo taken from within the flight deck) tested new spacesuits for shuttle crews. Unlike previous suits, these fully reusable designs could accommodate the somewhat variable dimensions of different wearers.

STS-7 *Challenger*

Launched	June 18, 1983
Landed	June 24, 1983
Commander	Robert L. Crippen
Pilot	Frederick H. Hauck

Mission specialists
John M. Fabian, Sally K. Ride, Norman E. Thagard

Milestones
On this mission, Ride became the first American woman to fly in space. Two communications satellites were deployed, ANIK C-2 for TELESAT Canada and PALAPA-B1 for Indonesia, both attached to small payload assist rocket motors. Seven Getaway Special canisters in the cargo bay held a variety of experiments, including one to study the effects of microgravity on an ant colony, while the joint U.S.-German Shuttle Pallet Satellite SPAS-01, a little platform crammed with ten scientific experiments, was deployed on the end of *Challenger*'s robot arm.

A superb portrait of Sally Ride, America's first woman in space, on the flight deck during the STS-7 mission.
▼

STS-8 *Challenger*

STS-8 crew member Guion Bluford using an exercise machine on *Challenger*'s *mid-deck.*

Launched	August 30, 1983
Landed	September 5, 1983
Commander	Richard H. Truly
Pilot	Daniel C. Brandenstein

Mission specialists

Guion S. Bluford Jr., Dale A. Gardner,
William E. Thornton

Milestones

Bluford became the first African-American to fly in
space. INSAT-1B, a multipurpose satellite for India,
attached to its payload assist motor, was deployed
into space. *Challenger*'s nose was held away from
sunlight for fourteen hours straight, to test the
effects of extreme cold on the orbiter's insulation
tiles and other systems.

*Lightning surges around the launch pad as STS-8 undergoes
final preparations. The tall spike on the gantry is just one of
many devices that eliminate the risk of lightning damage.*

STS-9 *Columbia*

Launched	November 28, 1983
Landed	December 8, 1983
Commander	John W. Young
Pilot	Brewster H. Shaw Jr.

Mission specialists

Owen K. Garriott, Robert A. Parker

Payload specialists

Byron K. Lichtenberg, Ulf Merbold

Milestones

For the first time in history, six people flew into space aboard a single spacecraft. One of them, German-born Merbold, represented the European Space Agency (ESA), builders of the substantial payload in *Challenger*'s cargo bay: Spacelab, a self-contained reusable scientific laboratory. No cash ever changed hands between the two space agencies. Europe provided the hardware in return for occasional access to space for its astronauts. The flight had some technical dramas. During re-entry, two of *Columbia*'s computers dropped off-line within five minutes of each other and an internal power unit caught fire (although the damage was not apparent until several days after after touchdown).

▲
The six crewmembers of STS-9 position themselves in a star-like cluster in the aft end cone of Spacelab. Clockwise, beginning with Young at middle right, are Merbold, Garriott, Shaw, Lichtenberg, and Parker.

Working in the STS-9 Spacelab module are, from left to right, Parker, Lichtenberg, Skylab veteran Garriott, and ESA astronaut Merbold.
▼

This mission was cancelled because of delays to its secret Department of Defense (DoD) payload. From now until 1986, a new numbering system for shuttle missions highlighted some of the political and bureaucratic priorities afflicting NASA throughout the mid-1980s, arguably at the expense of clear, honest information and sound engineering principles. The next flight, STS-41B, would be the first identified by the new nomenclature. The first digit, 4, represented the financial accounting year in which the mission was originally supposed to launch. The second digit, 1, designated which launch site was to be used, 1 for the Kennedy Space Center, and 2 for Vandenberg Air Force Base (a site that, in the event, was never actually made ready for launches). Finally, the third character denoted the sequential position a launch was supposed to occupy in the schedule for that financial year.

STS-41B *Challenger*

Launched	February 3, 1984
Landed	February 11, 1984
Commander	Vance D. Brand
Pilot	Robert L. Gibson

Mission specialists
Bruce McCandless II, Ronald E. McNair, Robert L. Stewart

Milestones
The first untethered space walks were carried out by McCandless and Stewart, using the Manned Maneuvering Unit (MMU), a gas-jet powered backpack. The WESTAR-VI and PALAPA-B2 satellites were deployed, but payload assist rocket motor failures left them stranded in low Earth orbit. The German-built Shuttle Pallet Satellite (SPAS), originally flown on STS-7, became the first satellite to be refurbished and carried back into space. SPAS remained in the payload bay when the remote manipulator arm exhibited an unexpected reluctance to operate properly.

◄
An excellent view of Challenger STS-41B's fiery ascent, and the accompanying vapor trail. Most of the visible smoke comes from the solid rocket boosters.

McCandless uses a foot restraint on the end of the robot arm.
Behind him are the Orbital Maneuvering System pods.

McCandless is transported along the cargo bay by the robot
arm. The protective logo-studded cradle that earlier housed
the WESTAR VI satellite is at the left corner of the frame.

▲ ▶

McCandless uses the nitrogen-propelled hand-controlled Manned Maneuvering Unit (MMU), performing this phase of the EVA without being tethered to the shuttle.

STS-41C *Challenger*

Launched	April 6, 1984
Landed	April 13, 1984
Commander	Robert L. Crippen
Pilot	Francis R. Scobee

Mission specialists
Terry J. Hart, George D. Nelson,
James Van Hoften

Milestones

Using an MMV and completing the job with *Challenger*'s robot arm, Nelson and Van Hoften captured the Solar Max satellite, replaced its attitude control system, then redeployed it into space, while a large-format IMAX movie camera captured the action. In addition, the Long Duration Exposure Facility (LDEF) was deployed and left in orbit to be retrieved by a future shuttle mission. LDEF's purpose was to test materials that might be suitable for a long life in orbit.

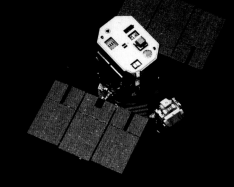

Solar Max, healthy once again after repairs to its gyro control system, is redeployed into space.

LDEF sets off for its long, lonely sojourn in orbit. Its range of materials is obvious from its patchwork appearance.

STS-41C's crew captured this dramatic photo of storm clouds gathering over South America. At lower right, the Paraná River snakes its way through the Brazilian rain forest.

STS-41D *Discovery*

Launched   August 30, 1984
Landed   September 5, 1984
Commander   Henry W. Hartsfield Jr.
Pilot   Michael L. Coats

Mission specialists
Steven A. Hawley, Richard M. Mullane,
Judith A. Resnick

Payload specialist
Charles D. Walker

Milestones

Three satellites were deployed during this mission: Satellite Business System SBS-D, SYNCOM IV-2 (also known as LEASAT2), and TELSTAR. The 102-foot-tall, 13-foot-wide Office of Application and Space Technology (OAST-1) solar wing was extended several times from the payload bay to demonstrate large, lightweight solar arrays for space stations. An IMAX camera was flown for a second time, along with an Air Force surveillance experiment, the futuristic-sounding Cloud Logic to Optimize Use of Defense Systems (CLOUDS).

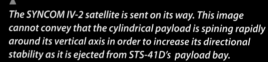

The SYNCOM IV-2 satellite is sent on its way. This image cannot convey that the cylindrical payload is spining rapidly around its vertical axis in order to increase its directional stability as it is ejected from STS-41D's payload bay.

▲
STS-41G crew member Sullivan on the flight deck with a pair of binoculars, prior to her history-making space walk.

▶
Sullivan is in center frame of this photo, with her fellow space walker Leestma at the right edge.

STS-41G *Challenger*

Launched   October 5, 1984
Landed   October 13, 1984
Commander   Robert L. Crippen
Pilot   Jon A. McBride

Mission specialists
David C. Leestma, Sally K. Ride,
Kathryn D. Sullivan

Payload specialists
Marc Garneau, Paul D. Scully-Power

Milestones

Sullivan became the first American woman to walk in space. The Earth Radiation Budget Satellite (ERBS) was deployed, and components of an Orbital Refueling System (ORS) tested the possibility of refueling spacecraft in orbit: a long-held dream of space planners. An IMAX camera was aboard for the third time, with a package of Canadian Experiments (CANEX), an atmospheric Auroral Photography Experiment (APE) and eight getaway specials: small, low-cost science payloads for a variety of users who might not otherwise have had access to space.

STS-51A *Discovery*

Launched   November 8, 1984
Landed   November 16, 1984
Commander   Frederick H. Hauck
Pilot   David M. Walker

Mission specialists
Joseph P. Allen, Anna L. Fisher, Dale A. Gardner

Milestones
The Canadian comsat TELESAT-H (ANIK) was deployed into geosynchronous orbit. The defense communications satellite SYNCOM IV-I (also known as LEASAT-1) was deployed. In a spectacular and headline-grabbing rescue operation, Allen and Gardner, wearing MMUs, retrieved two satellites, PALAPA-B2 and WESTAR-VI, both of which had been deployed during STS-41B in February that same year. Fisher then operated the remote manipulator system, grappling the satellites and pulling them securely into the payload bay.

As if in a scene from a science fiction movie, Gardner propels himself up to the drifting WESTAR satellite, and captures it with a grappling tool.

The shuttle system and its astronauts are riding high. Gardner clowns around with a "For Sale" sign after the successful retrieval of the two "used" satellites.

◄ *A wider shot of Gardner holding the "For Sale" sign, with Allen at right, standing on the Foot Restraint of the robot arm, and celebrating STS-51A's success.*

STS-51D *Discovery*

Launched	April 12, 1985
Landed	April 19, 1985
Commander	Karol J. Bobko
Pilot	Donald E. Williams

Mission specialists
S. David Griggs, Jeffrey A. Hoffman, M. Rhea Seddon

Payload specialists
Senator Jake Garn, Charles D. Walker

Milestones
The TELESAT-l (ANIK C-1) communications satellite was deployed, atop a payload assist module (PAM-D) rocket motor. SYNCOM IV-3 (also known as LEASAT-3) was also deployed but failed to do anything except drift about in *Discovery's* vicinity. Griggs and Hoffman performed an unscheduled space walk to try to nudge an actuating switch that might have woken the satellite from its unintended slumber, but nothing helped. SYNCOM was left to drift, not functioning, in low orbit, while mission planners looked at future options for retrieving it during another shuttle mission. The news media concentrated on the presence on board of U.S. Senator Jake Garn, taking part because of his boundless enthusiasm for space education and gamely trying to keep a smile on his face despite terrible nausea. Walker, an employee of the McDonnell Douglas Corporation rather than NASA, conducted biochemistry experiments as the world's first privately funded space traveler. A tire on *Discovery's* nosewheel assembly blew out on landing, but the craft was not endangered.

STS-51C *Discovery*

Launched	January 24, 1985
Landed	January 27, 1985
Commander	Thomas K. Mattingly
Pilot	Loren J. Shriver

Mission specialists
James F. Buchli, Ellison S. Onizuka

Payload specialist
Gary E. Payton

Milestones
This was a classified shuttle mission dedicated to the Department of Defense and the U.S. Air Force. A small Inertial Upper Stage (IUS) "kicker" rocket stage emerged from the shuttle's payload bay, carrying an Orion signals intelligence satellite. Parking itself in geosynchronous orbit, it unfurled a dish almost as wide as a football field to eavesdrop on ground communications and telemetry. Apparently, Orion met all its mission objectives. Even today, publicly available details about this payload remain scarce.

STS-51D deploys TELESAT to provide voice and TV coverage across Canada.
▼

▲
STS-51B commander Overmeyer might wish he could just stay on his flight deck, monitoring his instruments, but he still has to take his turn on the exercise machine.

STS-51B *Challenger*

Launched	April 29, 1985
Landed	May 6, 1985
Commander	Robert F. Overmyer
Pilot	Frederick D. Gregory

Mission specialists
Don L. Lind, Norman E. Thagard,
William E. Thornton

Payload specialists
Lodewijk van den Berg, Taylor G. Wang

Milestones
This was the second flight for the Spacelab orbital laboratory developed by the European Space Agency (ESA). Its mission was to provide a high-quality, shirtsleeve microgravity environment for delicate materials processing and fluid experiments. In addition to the human crew, two monkeys and two dozen rodents were monitored to see how they responded to weightlessness.

STS-51G *Discovery*

Launched	June 17, 1985
Landed	June 24, 1985
Commander	Daniel C. Brandenstein
Pilot	John O. Creighton

Mission specialists
John M. Fabian, Shannon W. Lucid,
Steven R. Nagel

Payload specialists
Patrick Baudry, Sultan Salman Al-Saud

Milestones
Three communications satellites, all attached to payload assist module motors, were deployed: MORELOS-A, for Mexico, ARABSAT-A, for the Arab Satellite Communications Organization and TELSTAR-3D for the AT&T company, along with a deployable and retrievable Shuttle Pointed Autonomous Research Tool for Astronomy (SPARTAN-1), six getaway special canisters, a Strategic Defense Initiative (SDI) experiment called the High Precision Tracking Experiment (HPTE), a materials processing furnace called Automated Directional Solidification Furnace (ADSF), and two French biomedical experiments. Saudi prince Al-Saud was on board to see his nation's payload safely away. He was the first Arab, the first Muslim, and the first member of royalty in space.

Discovery is just seconds away from a perfect touchdown at the conclusion of the STS-51G flight.
▼

The Instrument Pointing System for Spacelab-2 aims
the devices more accurately and kept them fixed on
their targets automatically.

STS-51F *Challenger*

Launched	July 29, 1985
Landed	August 6, 1985
Commander	C. Gordon Fullerton
Pilot	Roy D. Bridges Jr.

Mission specialists
Anthony W. England, Karl G. Henize,
F. Story Musgrave

Payload specialists
Loren W. Acton, John-David F. Bartoe

Milestones

Challenger's launch was unsettling. Three minutes and thirty seconds into the ascent, a cascade of temperature sensor failures led to the shutdown of the center engine: the only in-flight main engine failure during the entire shuttle program. *Challenger* made it safely to orbit. The Spacelab payload included three exposed pallets with scientific instruments dedicated to life sciences, plasma physics, astronomy, high-energy astrophysics, solar physics, atmospheric physics, and technology research. Additional instruments measured *Challenger*'s own infinitesimal electromagnetic and physical effects on the space around it. Down on Earth, people eagerly awaited vital data from the carbonated beverage dispenser evaluation, an experiment where both Coca-Cola and Pepsi tried to make their drinks palatable in space. Both seemed uncomfortably gassy in the shuttle's low cabin pressure.

STS-51I *Discovery*

Launched	August 27, 1985
Landed	September 3, 1985
Commander	Joe H. Engle
Pilot	Richard O. Covey

Mission specialists
William F. Fisher, James D. A. van Hoften,
John M. Lounge

Milestones
Three communications satellites were deployed:
ASC-1, for the American Satellite Company,
AUSSAT-1, an Australian communications satellite,
and SYNCOM IV-4 for the U.S. Navy. ASC-1 and
AUSSAT-1 were both attached to payload assist
motors. SYNCOM IV-4 made it to the correct
geosynchronous orbit but failed to function
thereafter. Fisher and van Hoften performed two
EVAs—retrieving, repairing, and redeploying the
SYNCOM IV-3, which had been deployed on
Discovery's 51D mission in April, but had been left
stranded in low Earth orbit.

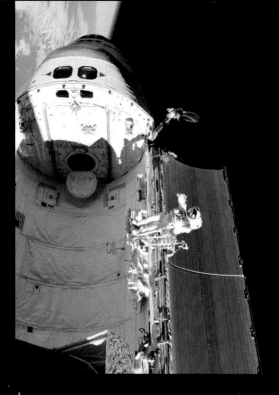

▲
*Fisher, anchored to a foot restraint on the starboard side
of* Discovery, *during STS-51I's capture and repair of the
Syncom IV-3 satellite.*

STS-51J *Atlantis*

Launched	October 3, 1985
Landed	October 7, 1985
Commander	Karol J. Bobko
Pilot	Ronald J. Grabe

Mission specialists
David C. Hilmers, William A. Pailes,
Robert L. Stewart

This was the second shuttle mission dedicated to
the Department of Defense. Two Defense Satellite
Communications System (DSCS) payloads were
deployed and successfully boosted into high
geosynchronous orbit by a single inertial upper
stage rocket motor.

▲
*STS-51J's Defense-related satellite and its small kicker stage
emerge from the cargo bay.*

STS-61A *Challenger*

Launched	October 30, 1985
Landed	November 6, 1985
Commander	Henry W. Hartsfield Jr.
Pilot	Steven R. Nagel

Mission specialists
Guion S. Bluford, James F. Buchli,
Bonnie J. Dunbar

Payload specialists
Reinhard Furrer, Ernst Messerschmid,
Wubbo J. Ockels

Milestones
This Spacelab mission featured a vestibular sled
designed to give scientists data on the functional
organization of human balance; in other words, it
was an experiment to reveal just how dizzy an
astronaut becomes when subjected to benevolent
torture at the hands of space medics. Astronaut-
scientists succumb to these procedures in the
interest of learning more about how to ensure
human health and survival during future long-term
space missions. While *Challenger* itself was
controlled as usual from NASA's Johnson Space
Center, scientific operations on Spacelab were run
by the German space operations center at
Oberpfaffenhofen, near Munich.

▲
*STS-61A's full crew. In the bottom row, from left to right, are
Messerschmid, Ockels, Nagel, and Bluford; in the back row,
Hartsfield, Dunbar, Buchli, and Furrer.*

STS-61B *Atlantis*

Launched   November 26, 1985
Landed   December 3, 1985
Commander   Brewster H.Shaw Jr.
Pilot   Bryan O'Connor

Mission specialists

Mary Cleave, Sherwood Spring, Jerry Ross

Payload specialists

Rodolfo Neri Vela, Charles D. Walker

Milestones

Three communications satellites were deployed: MORE LOS-B for Mexico, AUSSAT-2 for Australia, and SATCOM KU-2 for the RCA corporation. Two space walks by Spring and Ross tested assembly techniques for future space station applications. This photo shows Ross working on those tests.

The space community loves its acronyms. Ross works with EASE (Experimental Assembly of Structures in Extravehicular Activity), and ACCESS (Assembly Concept for Construction of Erectable Space Structure).

▼

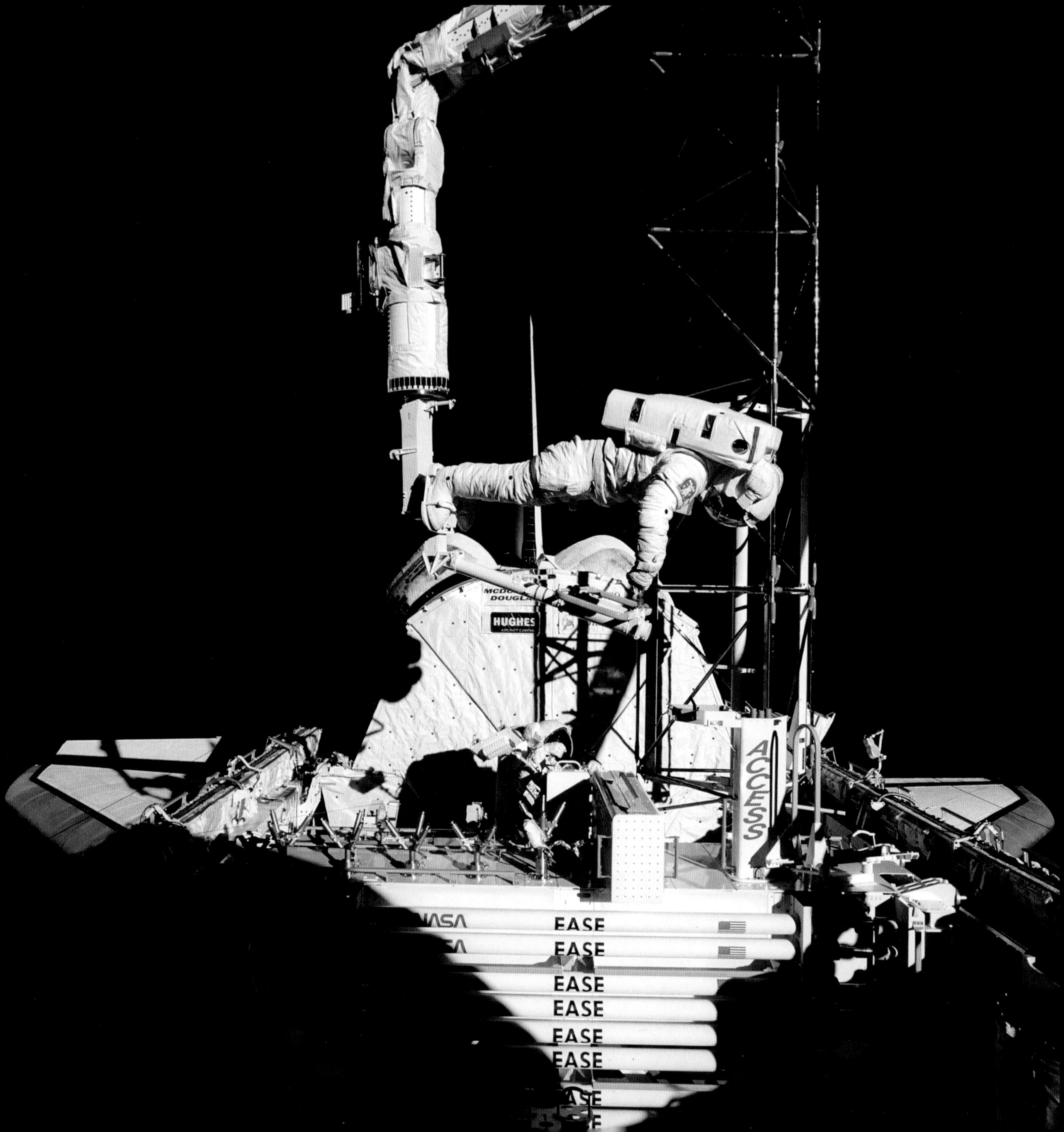

MCDO
DOUGL

HUGHES
AIRCRAFT COMPAN

ACCESS

NASA   EASE

NASA   EASE

EASE

EASE

EASE

EASE

At the commander's station of STS-61C, Gibson is in center frame, with future NASA Administrator Bolden to his right.

STS-61C *Columbia*

Launched	January 12, 1986
Landed	January 18, 1986
Commander	Robert L. Gibson
Pilot	Charles F. Bolden Jr.

Mission specialists

Franklin R. Chang-Diaz, Steven A. Hawley,
George D. Nelson

Milestones

The SATCOM KU-I (RCA Americom) satellite, attached to a payload assist motor, was deployed safely, but a camera that was supposed to monitor the Halley comet as it swept through the solar system on one of its rare visits ran into difficulties when its battery failed. A dozen getaway specials were mounted on a special platform in the payload bay. Congressman

said to him, back in 1986, that one day he would be NASA's chief, responsible for the tense and heartbreaking job of retiring the shuttle from service, and witnessing the layoffs of thousands of support staff, even though no replacement vehicle stood ready to take the shuttle's place.

Stage Two

FALLING
FROM GRACE
THE REALITIES OF ROCKET FLIGHT

FALLING FROM GRACE

THE REALITIES OF ROCKET FLIGHT

The world was shocked when the shuttle suddenly showed flaws in its design and in the way that it was operated by the people responsible for it. Behind every launch, there were perilous forces at work: shortage of money, a desire to please political masters, and a lack of true regard for safety. What had gone wrong with the agency that put men on the Moon?

NASA CELEBRATED ITS flagship spacecraft as "the most complex machine in history." In those first, innocent five years of orbital triumphs, few people outside the space industry realized that this fabulous complexity was not a virtue but a hazard. Judging by surface appearances, the shuttle was a science fiction dream come true, an awesome winged spaceship flying time and again into orbit and coming home to land like a plane. Nothing was more certain to turn dreams into nightmares than America's confidence that a machine powered by five furious infernos of hydrogen, oxygen, aluminum powder, and ammonium perchlorate could be considered so much tamed that its launches were barely worth remarking upon any more.

On the wintry late morning of January 28, 1986, STS-51L shuttle *Challenger* was launched from a pad gnarled with icicles into an uncharacteristically frigid Florida sky. The right-side solid rocket booster sprang a leak immediately on ignition. Freezing weather had compromised the rubber rings that sealed the joints between the booster's cylindrical segments. A jet of flame escaping from a faulty seal began to scorch the side of the huge liquid fuel tank to which the boosters and *Challenger* itself were attached. At seventy-three seconds after liftoff, the vehicle was 46,000 feet off the ground. Mission commander Dick Scobee got the all-clear from Mission Control to throttle up *Challenger*'s trio of liquid-fueled engines for the last push toward orbit. That was the moment when the wayward flame finally penetrated the skin of the tank, instantly igniting the potent fuels within. The tank detonated and STS-51L disintegrated at twice the speed of sound. Six astronauts and one civilian teacher, Christa McAuliffe, were killed.

"Flight controllers here looking very carefully at the situation," reported NASA's public affairs officer Steve Nesbitt in Houston, keeping as best he could to the facts, as was his proper task. "Obviously a major malfunction . . . We have no downlink." After a pause, he said, "We have a report from the flight dynamics officer that the vehicle has exploded."

Richard Feynman, a veteran of the 1940s Manhattan atomic bomb project and one of the most celebrated physicists of the twentieth century, was invited to join the presidentially mandated board of inquiry into *Challenger*'s loss: the Rogers Commission (chaired by former Secretary of State William P. Rogers). This independent-minded scientist immediately cleared his own eccentric path through the verbiage. For instance, with the TV cameras running, he dipped a piece of rubber into a glass of iced water and showed how it hardened when cold. "Do you suppose this might have some relevance to our problem?" he asked, knowing very well that it did. He had created a vivid demonstration of the notorious O-ring flaw that had destroyed the shuttle during its icy launch.

Asking awkward questions

Feynman wanted to probe further than the Rogers Commission had expected from him. "If NASA was slipshod about the leaking rubber seals on the solid rockets, what would we find if we looked at the liquid-fueled engines and all the other parts that make up a Shuttle?" He was told that it was way beyond the Commission's brief to look at the main engines because no problems had been reported. So he made unauthorized trips to NASA facilities where he could

The crew of STS-51L Challenger. *In the back row, from left to right: Ellison S. Onizuka, Sharon Christa McAuliffe, Greg Jarvis, and Judy Resnik. In the front row, from left to right: Michael J. Smith, Dick Scobee, and Ron McNair.*

▼

speak to ground-floor engineers in private. He wrote later, "I had the definite impression that senior managers were allowing errors that the Shuttle wasn't designed to cope with, while junior engineers were screaming for help and being ignored."

Feynman had identified a serious and endemic managerial problem within NASA. Feynman also looked at the complex relationships between the space agency's many departments and their private industrial suppliers. "NASA's propulsion office in Huntsville designs the engines, Rocketdyne in California builds them, Lockheed writes the instructions and the launch center

in Florida installs them. It may be a genius system of organization, but it seems a complete fuzzdazzle to me." In the last days of the inquiry, and in the final written report, he made a plea for greater realism. "For a successful technology, reality must take precedence over public relations, because Nature cannot be fooled."

Obviously there was more to the Shuttle disaster than hardware failures alone, but it took a special kind of expert to suggest exactly what those factors might be. Diane Vaughan, Professor of Sociology and Public Affairs at Columbia University, New York, spent nine years collating data for her landmark study of the 1986

▶

At the lower right of this image, taken just seconds after ignition of the solid rocket boosters, a tell-tale puff of dark smoke can be seen, indicating an improper escape of hot gas from the starboard booster.

accident. The Challenger Launch Decision: Risky Technology, Culture, and Deviance at NASA was published to widespread acclaim on the tenth anniversary of the explosion. The common perception is that NASA managers must have been reckless to launch a shuttle on such an unexpectedly cold day. Somewhat to her surprise, Vaughan found that this was not so. "After looking more closely at the data, it turned out that the managers had not violated rules at all, but had actually conformed to all NASA's requirements. In their view, they were obeying the correct engineering and organizational principles."

Officials responsible for the launch had acted in good faith. And yet, the system had failed. Worse still, the formal investigation found that O-rings had suffered partial burn-throughs on at least five previous shuttle missions. As Vaughan explains, NASA "repeatedly observed the problem with no consequence and reached the point where flying with a known flaw was normal and acceptable. Of course, after the accident, they were shocked and horrified when they saw what they had done."

WAS NASA CARELESS?

The story handed down to us—the judgment of received wisdom, so to speak—was that *Challenger* was destroyed as a result of careless decisions, especially in relation to the O-rings. Within NASA, the Rogers Commission identified "incomplete and sometimes misleading information, a conflict between engineering data and management judgments, and a management structure that permitted internal flight safety problems to bypass key shuttle managers." However, according to Roger Launius, formerly NASA's chief historian and now chief space curator at the Smithsonian Air and Space Museum in Washington, D.C., "detailed analysis of documentary evidence and testimony shows that NASA personnel involved in the O-ring question were convinced that the joints were safe, and that there were numerous other problems—especially with the shuttle main engines, and the fuel turbine pumps—that consumed most of their attention."

Unfortunately, that confidence was based on reports from the O-ring manufacturer,

Morton Thiokol, which had tried to downplay the concerns of one its own engineers, Roger Boisjoly, about the seals' potentially poor performance in cold weather. Boisjoly had raised the issue a year prior to *Challenger*'s last flight, and he, along with several other engineers, raised it again the day before the ill-fated launch. Although a telephone conference took place between NASA and Morton Thiokol, the Thiokol managers backpedaled from Boisjoly's warnings. NASA managers, eager to fly the mission after several previous delays, put aside what they saw as inconclusive concerns, and went ahead with the launch.

Perhaps the people best qualified to judge NASA's performance were the astronauts who took the actual risks of flying the shuttle. NASA's most experienced commander, John Young, veteran of Gemini, Apollo, and the shuttle's inaugural flight to orbit, wrote a blistering internal memo, echoing the thoughts of many colleagues. "If we do not consider flight safety first, all the time, at all levels of NASA, this machinery and this program will not make it. If the management system is not big enough to stop the space shuttle program whenever necessary to make flight safety corrections, it will not survive, and neither will our three space shuttles or their flight crews."

However, by 1988 the space agency seemed to be turning the corner. Sally Ride, America's first woman in space, had seen the very orbiter in which she had traveled blown apart. Now, as a contributor to the Rogers Commission, she thought she could detect signs of hope. "NASA has put in a lot of time to improve the safety of the space shuttle to fix the things that had gone wrong. The shuttle is a better and safer rocket than it was before the accident."

No one wanted this ship to explode. All concerned worked with the safety of the astronauts as their priority. And yet, things went wrong. We are only human.

Pre-launch inspection photos of STS-51L Challenger on the launch pad showed icicles hanging from the surrounding gantry structures. In retrospect it seems impossible that anyone could have allowed the launch to proceed, given the harsh weather conditions. Common sense took a back seat that day, but this was more the fault of NASA's flawed bureaucratic structures than its individual people.

Wreckage and boxes of debris from Challenger are lowered into an abandoned Minuteman Missile silo at Complex 31 on Cape Canaveral Air Force Station for permanent burial.

REBUILDING
TRUST

AN ERA OF GREAT ACHIEVEMENTS

Stage Three

REBUILDING TRUST
AN ERA OF GREAT ACHIEVEMENTS

As Americans wondered if flying the shuttles was worth the risks, the vehicle at last began to show off what it could achieve. A new and productive alliance between skilled humans and complex machines took hold in orbit.

FOR OUR GENERATION, the *Challenger* catastrophe defined NASA's public image both as the architect of the Apollo 11 triumph and as a complicated, flawed agency—but it can always be counted upon to come back from disaster and flash its old-time "can-do" smile. Once the shuttle had returned to flight, a particular mission caught the public imagination. It was widely hailed as among the most worthwhile projects ever undertaken by astronauts.

After its launch in April 1990 from the payload bay of shuttle *Endeavour*, the long-awaited Hubble Space Telescope began its operational career in disarray. Its main mirror was manufactured to the wrong specification. The telescope that was supposed to give us our sharpest-ever view of the farthest reaches of space was essentially blind. Just as embarrassing, the huge solar panels that powered the on-board equipment flexed and buckled every time Hubble's orbit took it into, or out of, direct exposure to the sun, shaking the telescope so much as to make it virtually worthless as a precision instrument. The $2-billion machine, a decade in development, was apparently a farce. Fortunately, the telescope had been designed for future in-orbit servicing by shuttle astronauts. Optical engineers realized that the faulty mirror could be corrected at the "eyepiece" end of the assembly, using a clever system of lenses to bend the fuzzy images back into focus. Meanwhile, the British company that had manufactured the solar panels worked out why they were behaving so poorly and prepared a new set of panels that would remain stable during temperature fluctuations when the telescope passed between cold shadow and glaring sunlight during its orbit.

MAGNIFICENT REPAIR

In December 1993, the Hubble Space Telescope was pulled into *Endeavour*'s cargo bay using the Canadian-built remote manipulator robot arm. Musgrave and Hoffman secured Hubble and began replacing equipment. Then Thornton used a pair of shears to snip off the old solar panels, while *Endeavour* produced a brief burst of thruster fire to send the panels safely down into the atmosphere.

Thornton and Akers then loaded a self-contained package of corrective optics into a side hatch on Hubble. During subsequent space walks, the crew added more instruments and new gyroscope control systems before gently easing Hubble free of the *Endeavour* and back into an independent orbit. Hubble's instruments may have been flawed, but the main spacecraft structure turned out to have been well designed. The access doors opened properly, and the new equipment slid into place. The space walks were spectacular, and NASA's renewed confidence set an optimistic tone ahead of the International Space Station project. This image shows a stellar "nursery" in the Orion nebula: one of countless images returned by the telescope since its repair. The public has never questioned the merits of this project.

Regaining confidence

On December 2, 1993, shuttle *Endeavour* lifted off carrying a rescue crew for the Hubble. Working as two separate duty teams, astronauts Story Musgrave, Jeffrey Hoffman, Thomas Akers, and Kathryn Thornton completed the necessary work in five back-to-back space walks totaling thirty-five hours. Many of NASA's critics began to revise their views, perhaps unwilling to appear ungracious in the light of an obvious success. The complex space walks also demonstrated that establishing a permanent human presence in space need not be as foolhardy an idea as some commentators had imagined.

Everybody at NASA had always imagined the shuttle would be just one element of that old *Collier's* dream, an orbiting infrastructure. What was the point of a shuttle if it had no proper destination to reach? Many detailed proposals had been formulated, but no serious budget requests were put forward. The space agency gambled on proving their shuttle concept first and then winning funds for more elements in future years. The Hubble repair mission reawakened public and political enthusiasm for astronaut adventures. It also proved that humans were not merely capable of performing complex assembly tasks up there. They were essential for any broad-reaching space program.

The International Element

Just a few months before Russia's Sputnik was launched, Europe made its own leap into the future. On March 25, 1957, the Treaty of Rome gave birth to the European Economic Community (EEC). The political ideals behind the treaty were reflected in many areas of social, economic, and cultural collaborations between member states, including the joint development of new science and technology projects. For instance, in that same year, 1957, the European Organization for Nuclear Research, commonly known as CERN, activated its first subatomic particle accelerator experiments. By 1958 a number of prominent European scientists were discussing if a similar organizational approach could be applied to space research. In the early 1960s, two agencies, the European Launch Development Organization (ELDO) and the European Space Research Organization (ESRO), were created, the first, to build rockets and the second, to run missions. At that time there were no obvious plans to create a European human presence in space. This was about to change.

The final Apollo lunar landing mission was completed in December 1972, by which point the space race with the Soviets was no longer considered an urgent issue. As development of the shuttle progressed,

NASA and the White House alike began to think less about rivalry with foreign nations and looked, instead, toward greater cooperation with its allies, and especially Western Europe. In the early 1970s, as ESRO and ELDO began to unify into a single European Space Agency (ESA), the possibility emerged that European companies might build some components for the shuttle, such as the tail plane or the 60-foot-long payload bay doors. A second option was to design an intermediate transfer stage or "space tug" that would help boost shuttle payloads into a higher orbit once they were released from the cargo bay. None of these ideas were exactly right for ESA, and when the crunch came, no one at NASA or in Washington was keen to allow parts of the shuttle's hardware to become dependent on foreign manufacturers a long way overseas. Everyone concluded that European industry should design and build a separate payload that could fly inside the shuttle's cargo bay, yet not be an integral part of it.

Spacelab was born as a distinctly European space system, even if it called upon NASA's launch capability in order to reach orbit. Spacelab was a modular system with a pressurized crew compartment and an exchangeable series of nonpressurized external pallets carrying experiments that required raw exposure to the space environment. A pressurized connecting tunnel linked Spacelab with the mid-deck of the shuttle's crew compartment. Spacelab was essentially a complete miniature space station, an ambitious payload that would have been difficult for NASA to fund and build at that time because the available resources were fully

stretched by the development of the shuttle itself. Similarly, it was not realistic at that time for Europe to launch a large payload such as Spacelab without access to the shuttle. The European-American partnership created a joint space project that was more ambitious than either side could have achieved on its own. The barter arrangements in which launch services and access to space were exchanged for European hardware limited the need for cash transfers between either party and established an arrangement whose benefits are still in place today. The Spacelab decision was the moment when human spaceflight adventure became truly international and collaborative.

Spacelab flew for the first time on November 28, 1983, aboard *Columbia*. One of the three ESA astronauts trained for the mission, the German citizen Ulf Merbold, was the first non-U.S. citizen to fly on the shuttle and the second European citizen to fly in space (the first being Sigmund Jaehn, on a Russian Soyuz spacecraft in August 1978.) In April 1985, Spacelab was in orbit once again, this time aboard *Challenger*. Spacelab components flew on a total of twenty-two missions, with sixteen featuring the pressurized module. The project supported research from more than two hundred scientists from thirteen countries. The longest Spacelab mission, launched in June 1996, lasted for seventeen days. A few months after that mission, a new era began. Humans were about to occupy Earth orbit permanently, rather than just for a few days at a time.

MISSION IMAGES
DECEMBER 1988 – NOVEMBER 1994

The launch of STS-26 restores the shuttle to flight status
after a distressing two years of self-reflection for NASA.

STS-26 *Discovery*

Launched	September 29, 1988
Landed	October 3, 1988
Commander	Frederick H. Hauck
Pilot	Richard O. Covey

Mission specialists
David C. Hilmers, John M. Lounge,
George D. Nelson

Milestones
The primary payload, NASA's Tracking and Data
Relay Satellite-3 (TDRS-3), was attached to an
Inertial Upper Stage (IUS) rocket and deployed
and boosted to geosynchronous orbit. Everyone
was relieved to see the shuttle system flying again
after a traumatic two years and eight months of
personal and national grief, and major technical
overhauls of the shuttle hardware.

◄

A spectcacular view of Atlantis *ascending at the start of
STS-27. Notice that the three hydrogen and oxygen-fueled
main engines create very little smoke. Their exhaust product
is pure superheated water vapor.*

STS-27 *Atlantis*

Launched	December 2, 1988
Landed	December 6, 1988
Commander	Robert L. Gibson
Pilot	Guy S. Gardner

Mission specialists
Richard Mullane, Jerry L. Ross,
William M. Shepherd

Milestones
This mission, dedicated to the Department of
Defense, was so secret that even the exact timing
of the launch was not announced in advance, just
in case hostile foreign powers might glean
something of the payload's nature from its
potential orbital charateristics. The Lacrosse radar
reconnaissance satellite failed after deployment.
Atlantis was nudged alongside it. The astronauts
completed repairs and sent it on its way.

STS-29 *Discovery*

Launched   March 13, 1989
Landed   March 18, 1989
Commander   Michael L. Coats
Pilot   John E. Blaha

Mission specialists
James P. Bagian, James F. Buchli,
Robert C. Springer

Milestones
The primary payload was another in a gradually expanding fleet of tracking and data relay satellites, attached to an IUS. After deployment, it successfully reached geosynchronous orbit. *Discovery*'s on-board workload hinted at the grand space project soon to come. The Space Station Heat Pipe Advanced Radiator Experiment (SHARE) was tested, ahead of the impending International Space Station (ISS) assembly missions. *Discovery*'s crew also photographed Earth with a hand-held IMAX camera.

▶
STS-29's empty External Tank, as tall as a fifteen-story building, but useless once depleted, falls back towards a watery grave at the bottom of the Atlantic Ocean.

STS-30 *Atlantis*

Launched   May 4, 1989
Landed   May 8, 1989
Commander   David M. Walker
Pilot   Ronald J. Grabe

Mission specialists
Mary L. Cleave, Mark C. Lee, Norman E. Thagard

Milestones
The primary payload was Magellan, an ambitious radar mapping spacecraft destined for the planet Venus. After deployment from the payload bay, Magellan's first and second stage rocket motors fired as planned, boosting the spacecraft onto its proper trajectory for the fifteen-month journey to Venus, where it accomplished all its mission goals.

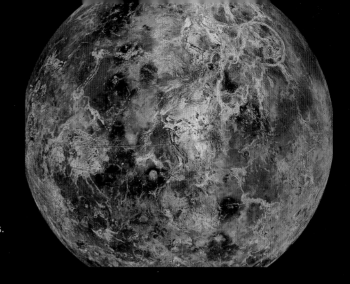

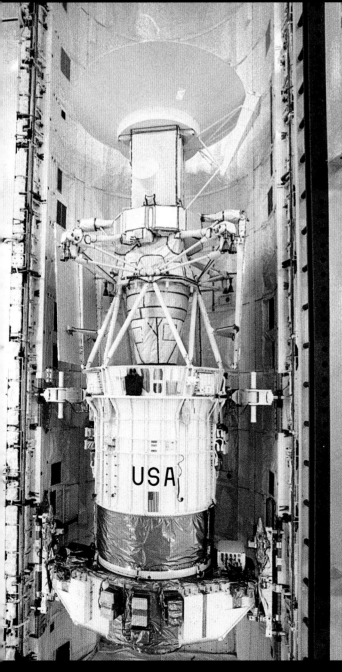

Magellan in STS-30's cargo bay prior to launch, and (at lower right) emerging into space just before final deployment.

Magellan's cloud-penetrating radar scan of the Venusian surface, where the temperature is hot enough to melt lead.

Launched	August 8, 1989
Landed	August 13, 1989
Commander	Brewster H. Shaw Jr.
Pilot	Richard N. Richards

Mission specialists

James C. Adamson, Mark N. Brown,
David C. Leestma

Milestones

This was the fourth shuttle mission dedicated
entirely to the Department of Defense. The
principal payload was a military Satellite Data
System (SDS) communications satellite. Also on
board, among several scientific experiments, was
a unique object managed jointly by NASA and the
DoD, and known at the time only as "Detailed
Secondary Objective 469." This was a human skull
(female), encased in plastic, testing the extent to
which space radiation penetrates the body.

*...landlocked and exceptionally salty body of water is more
than a thousand feet below normal sea level.*

*Galileo in a special clean room at the Kennedy Space Center,
undergoing final checks before its installation in the cargo
bay of its delivery vehicle, STS-34 Atlantis.*

STS-34 *Atlantis*

Launched	October 18, 1989
Landed	October 23, 1989
Commander	Donald E. Williams
Pilot	Michael J. McCulley

Mission specialists

Ellen S. Baker, Franklin R. Chang-Diaz,
Shannon W. Lucid

Milestones

The Galileo spacecraft, destined for Jupiter, was
successfully deployed six hours into the flight. The
upper stages fired correctly, putting Galileo on
course for its epic six-year trip to Jupiter, via
gravitational boosts from Venus and a second
swing-by of Earth to pick up yet more velocity
before it headed into the deep reaches of the
solar system. Despite the partial unfurling of its
main antenna, Galileo was a brilliant success.

STS-33 *Discovery*

Launched	November 22, 1989
Landed	November 27, 1989
Commander	Frederick D. Gregory
Pilot	John E. Blaha

Mission specialists
Manley L. Carter Jr., F. Story Musgrave,
Kathryn C. Thornton

Milestones
A fifth mission dedicated to the Department of
Defense, carrying a Magnum "Big Ear" ELINT
(ELectronic INTtelligence) satellite. Gregory was
the first African-American spacecraft commander.

▶

*A spectacular, almost blindingly bright night launch
for Columbia at the start of mission STS-33.*

LDEF patiently waits for STS-32 Columbia to grapple it with the robot arm and haul it back into the cargo bay for return to Earth after more than four years adrift in space. As the arm pulls LDEF ever closer toward the payload bay, discoloration on some of its test panels becomes visually evident.

STS-32 *Columbia*

Launched	January 9, 1990
Landed	January 20, 1990
Commander	Daniel C. Brandenstein
Pilot	James D. Wetherbee

Mission specialists
Bonnie J. Dunbar, Marsha S. Ivins, G. David Low

Milestones
The prime objective was the deployment of SYNCOM IV-F5 defense communications satellite and retrieval of NASA's Long Duration Exposure Facility (LDEF), left in orbit by STS-41C *Challenger* back in April 1984. The retrieval had been delayed for more than four years because of scheduling changes after *Challenger*'s catastrophic loss in 1986. LDEF's successful capture was filmed with an IMAX camera, and footage subsequently appeared in the ultra large-screen theatrical presentations *Blue Planet* and *Destiny in Space*.

STS-36 *Atlantis*

Launched	February 28, 1990
Landed	March 4, 1990
Commander	John O. Creighton
Pilot	John H. Casper

Mission specialists
David C. Hilmers, Richard M. Mullane,
Pierre J. Thuot

Milestones

The sixth shuttle mission dedicated to the
Department of Defense, based on deployment of
a stealth reconnaissance satellite code-named
Misty. Its task may have been to monitor Soviet
weapons development, even though, by now, the
Soviet era was nearing its end.

▲
*Hubble's solar panels unfurl like golden roller blinds while
the telescope is still safely attached to STS-31's robot arm.*

STS-31 *Discovery*

Launched	April 24, 1990
Landed	April 29, 1990
Commander	Loren J. Shriver
Pilot	Charles F. Bolden Jr.

Mission specialists
Steven A. Hawley, Bruce McCandless,
Kathryn D. Sullivan

Milestones

This mission successfully deployed the famous
Hubble Space Telescope, while an IMAX camera
documented the procedure. It soon became
apparent that the telescope's primary optics were
flawed, and it was all but useless. It would have to
be salvaged, somehow, by a future shuttle
mission. What seemed, at first, like an
embarassing letdown turned out to be one of the
most productive alliances between humans and
machines that our culture has ever known. With
help from astronaut repair teams during the 1990s
and the first decade of the twenty-first century,
Hubble emerged as the world's greatest and most
popular scientific instrument. It was regarded by
many commentators as a significant vindication of
the shuttle itself.

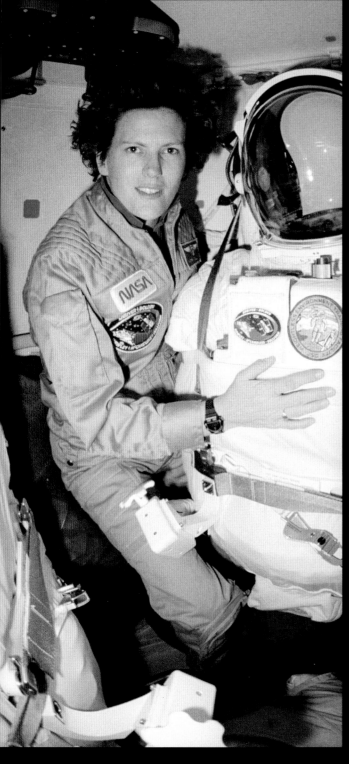

Sullivan poses in the airlock with her spacesuit, just in case Hubble's deployment needs a helping hand. All goes well and no EVA is required.

The moment of release: Hubble is set free from the end of the robot arm to begin its independent career in orbit. No one at the time knew what lay ahead: the flaws in the optical system, and the subsequent repair effort.

STS-41 *Discovery*

Launched   October 6, 1990
Landed   October 10, 1990
Commander   Richard N. Richards
Pilot   Robert D. Cabana

Mission specialists
Thomas D. Akers, William M. Shepherd,
Bruce E. Melnick

Milestones
The primary payload (and one of the heaviest so
far in the shuttle's history) was the ESA-built
Ulysses spacecraft, designed to explore the polar
regions of the sun. An IUS mated to a payload
assist module yielded a double kick of propulsion
to send Ulysses far out of the plane of the ecliptic,
the disc-shaped realm of space around which
most of the solar system's planets, moons, and
asteroids drift around the sun.

▲
*The massive Ulysses solar probe and its detachable rocket
motor are cast into space by STS-41.*

▲
*A beguiling STS-38 photo of golden sunlight, sparkling
sea and clouds on the most fascinating of all planets in
the solar system: our own world.*

STS-38 *Atlantis*

Launched   November 15, 1990
Landed   November 20, 1990
Commander   Richard O. Covey
Pilot   Frank L. Culbertson Jr.

Mission specialists
Charles D. Gemar, Carl J. Meade,
Robert C. Springer

Milestones
The seventh mission dedicated to Department of
Defense, deploying a reconnaissance satellite,
Magnum 3. As usual with such payloads, the
nature of its mission was secret, but it may have
monitored the Persian Gulf during this tense
period of U.S. military activity associated with the

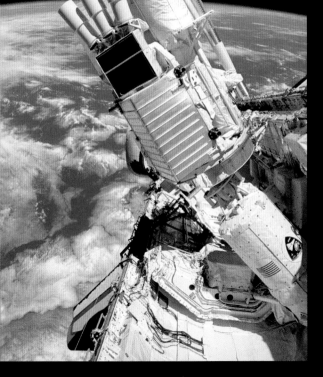

◄ *The ASTRO-1 payload of astronomical instruments protruding from the payoad bay during STS-35.*

STS-35 *Columbia*

Launched	December 2, 1990
Landed	December 10, 1990
Commander	Vance D. Brand
Pilot	Guy S. Gardner

Mission specialists
Jeffrey A. Hoffman, John M. Lounge,
Robert A. Parker

Payload specialists
Samuel T. Durrance, Ronald A. Parise

Milestones
The primary objectives were round-the-clock
astronomical observations in ultraviolet and x-ray
wavelengths. The ASTRO-1 payload consisted of
four instruments: the Hopkins Ultraviolet
Telescope (HUT), the Wisconsin Ultraviolet Photo-
Polarimeter Experiment (WUPPE), an Ultraviolet
Imaging Telescope (UIT), and a Broad Band X-Ray
Telescope (BBXRT). Ultraviolet telescopes
mounted on Spacelab elements in the cargo bay
were supposed to be operated in shifts by the
flight crew. Problems with the automatic electronic
pointing systems forced ground teams at NASA's
Marshall Space Flight Center to help the crew aim
some of the telescopes manually. However, most
of the science data was obtained successfully.

STS-37 *Atlantis*

Launched	April 5, 1991
Landed	April 11, 1991
Commander	Steven R. Nagel
Pilot	Kenneth D. Cameron

Mission specialists
Jay Apt, Linda M. Godwin, Jerry L. Ross

Milestones
The massive Gamma Ray Observatory (GRO) was
deployed on flight day three. Its main antenna
failed to deploy on command, but it was finally
freed and manually deployed by Ross and Apt
during an unscheduled space walk—yet another
example of the constructive union between human
space explorers and robotic machines. An
additional six-hour space walk by astronauts Ross
and Apt tested systems that would one day help
astronauts navigate themselves safely around the
exterior of the International Space Station.

▲ *The Gamma Ray Observatory on the end of the robot arm, moments before deployment into space by STS-37.*

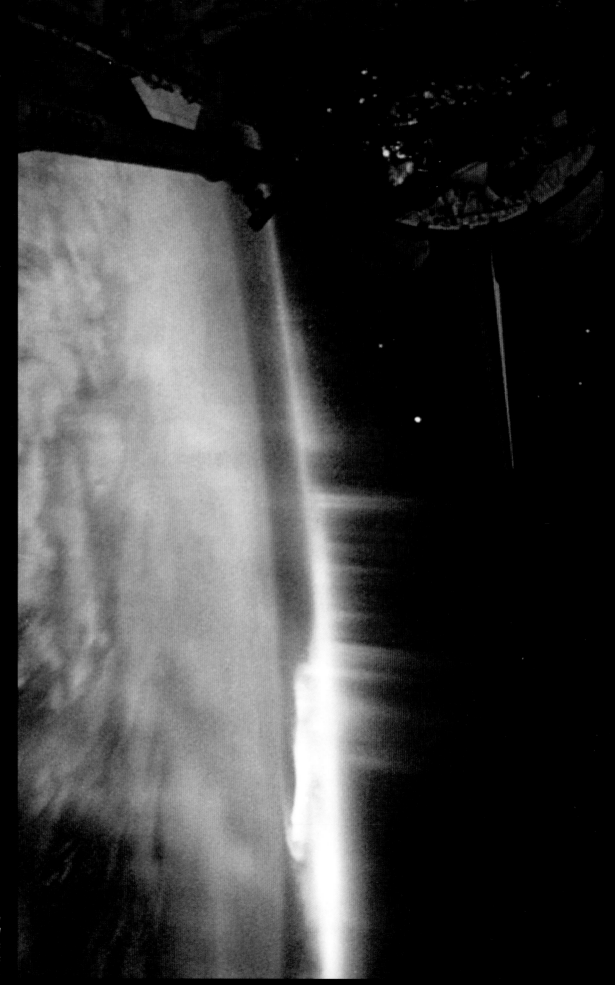

STS-39 *Discovery*

Launched	April 28, 1991
Landed	May 6, 1991
Commander	Michael L. Coats
Pilot	L. Blaine Hammond

Mission specialists
Guion S. Bluford Jr., Gregory J. Harbaugh,
Richard J. Hieb, Donald R. McMonagle,
Charles L. Veach

Milestones
An unclassified Department of Defense payload
included infrared instruments, radiation
monitoring equipment, and other experiments.
The unusually high orbital path chosen for this
mission, slanted at 57 degrees relative to the
equator, allowed *Discovery* to fly over most of the
Earth's major land masses at least once.

*Red and green colors predominate in this view of the Aurora
Australis photographed from STS-39. Auroras are caused
when high-energy electrons pour down from the Earth's
magnetosphere and collide with atoms. The eerie glow is
light emitted when the atoms return to their original state.*

Launched	June 5, 1991
Landed	June 14, 1991
Commander	Bryan D. O'Connor
Pilot	Sidney M. Gutierrez

Mission specialists

James P. Bagian, F. Drew Gaffney,
Millie Hughes-Fulford, Tamara E. Jernigan,
M. Rhea Seddon

Milestones

This Spacelab mission featured the most detailed physiological measurements in space since the Skylab program in the mid-1970s (Skylab was America's first space station, built around Apollo-Saturn hardware). The test subjects for this mission were the humans, plus thirty rodents and several thousand tiny jellyfish.

A great view of Columbia coming down to a perfect landing at the end of mission STS-40. This orbiter was destroyed during reentry on the first day of February 2003.

▼

STS-43 *Atlantis*

Launched	August 2, 1991
Landed	August 11, 1991
Commander	John E. Blaha
Pilot	Michael E. Baker

Mission specialists
James C. Adamson, G. David Low,
Shannon W. Lucid

Milestones
Tracking and Data Relay Satellite-5 (TDRS-5),
attached to an IUS, was deployed about six hours
into flight, and propelled into geosynchronous
orbit, where it joined a small constellation of
similar systems used by NASA and other U.S.
government agencies for communications with
various orbiting spacecraft.

◄
*The twin fury of the solid rocket boosters carries most of the
initial load as a fuel-laden STS-43 leaves the launch pad.*

A close-up view of the Upper Atmosphere Research Satellite eployed by STS-48. The gold-colored "Mylar" fabric is common to countless spacecraft. Though gossamer-thin and lightweight, it is an effective shield against the intense heat of the sun.

STS-48 *Discovery*

Launched	September 12, 1991
Landed	September 18, 1991
Commander	John O. Creighton
Pilot	Kenneth S. Reightler Jr.

Mission specialists

Mark N. Brown, James F. Buchli, Charles D. Gemar

Milestones

The Upper Atmosphere Research Satellite (UARS) was deployed on the third day of the mission, on its way toward an eighteen-month mission to conduct the most detailed study ever made of the Earth's troposphere, the upper level of the planet's breathable region of atmosphere.

STS-44 *Atlantis*

Launched   November 24, 1991
Landed   December 1, 1991
Commander   Frederick D. Gregory
Pilot   Terence T. Henricks

Mission specialists
Thomas J. Hennen, F. Story Musgrave,
Mario Runco Jr., James S. Voss

A dedicated Department of Defense mission,
including a Defense Support Program (DSP)
satellite and attached IUS deployed on flight day
one and a range of experiments in the cargo bay.

STS-42 *Discovery*

Launched   January 22, 1992
Landed   January 30, 1992
Commander   Ronald J. Grabe
Pilot   Stephen S. Oswald

Mission specialists
David C. Hilmers, William F. Readdy,
Norman E. Thagard

Payload specialists
Roberta L. Bondar, Ulf D. Merbold

Milestones
The International Microgravity Laboratory-1
(IML-1), made its first flight, using the pressurized
Spacelab module. The international crew was
divided into two teams for around-the-clock
research on the human nervous system's
adaptation to low gravity and the effects of
microgravity on other life forms, such as shrimp
eggs, lentil seedlings, fruit fly eggs, and bacteria.
Materials processing experiments were also
conducted, including crystal growth from a variety
of substances, including exotic mercury
compounds and biological enzymes.

▶

*STS-42 astronaut Hilmers endures the Microgravity
Vestibular Investigations experiment, sitting in its rotating
chair inside the IML-1 module. The device monitors the
effects of microgravity on our senses.*

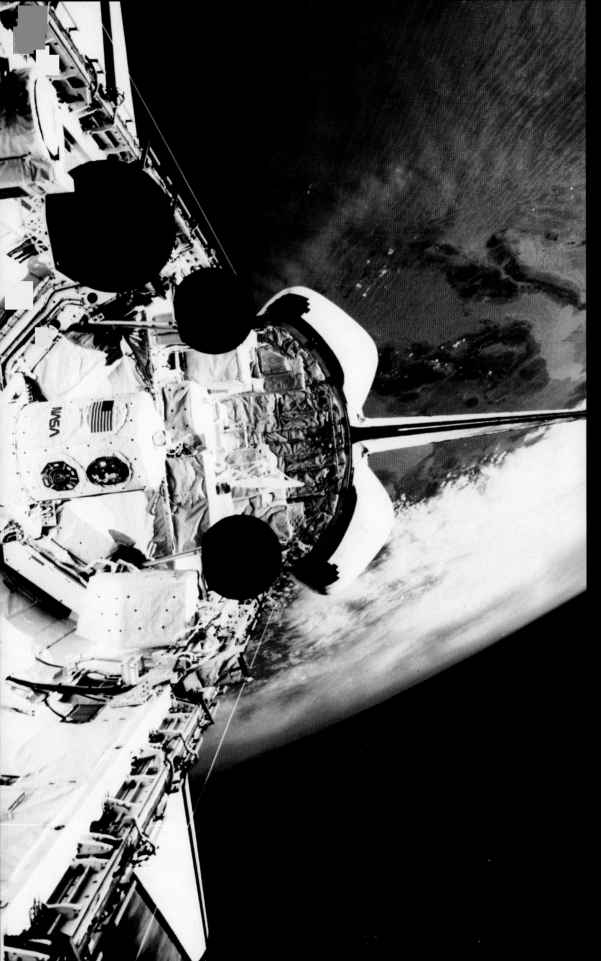

STS-45 *Atlantis*

Launched	March 24, 1992
Landed	April 2, 1992
Commander	Charles F. Bolden Jr.
Pilot	Brian Duffy

Mission specialists
C. Michael Foale, David C. Leestma,
Kathryn D. Sullivan

Payload specialists
Dirk D. Frimout, Byron K. Lichtenberg

Milestones

The mission carried the first Atmospheric
Laboratory for Applications and Science (ATLAS-1)
on Spacelab pallets mounted in *Atlantis*' cargo
bay. The nondeployable payload—equipped with
twelve instruments from the United States,
France, Germany, Belgium, Switzerland, the
Netherlands, and Japan—conducted studies in
atmospheric chemistry, solar radiation, space
plasma physics, and ultraviolet astronomy.

◄

What seem like dark discs in this image are actually
spheres of protective cladding around a series of particle
physics accelerator experiments within the ATLAS-1 suite.

The balky satellite awaits capture as STS-49 Endeavour makes a careful approach to its target.

▶

From right to left, STS-49 astronauts Hieb, Akers, and Thuot gain firm handholds on the satellite.

STS-49 *Endeavour*

Launched	May 7, 1992
Landed	May 16, 1992
Commander	Daniel C. Brandenstein
Pilot	Kevin P. Chilton

Mission specialists
Thomas D. Akers, Richard J. Hieb,
Bruce E. Melnick, Kathryn C. Thornton,
Pierre J. Thuot

Milestones
The INTELSAT VI (F-3) satellite, stranded in an
unusable orbit since its launch aboard a Titan
expendable launch vehicle in March 1990, was
captured by crewmembers, who then installed a
new rocket motor package. The satellite was
subsequently released into orbit, and the motor
successfully kicked the spacecraft into a
geosynchronous orbit for operational use. The
difficult capture required three EVAs by astronauts
Thuot and Hieb, who were at first unable to attach
a capture bar to the satellite. A second attempt
the following day was also unsuccessful. Finally,
Thuot and Hieb, now joined by Akers, simply
grabbed the reluctant INTELSAT by hand, while
Brandenstein delicately maneuvered the orbiter
to within a few feet of the four-ton satellite.

Landed	July 9, 1992
Commander	Richard N. Richards
Pilot	Kenneth D. Bowersox

Mission specialists
Ellen S. Baker, Bonnie J. Dunbar, Carl J. Meade

Payload specialists
Lawrence J. DeLucas, Eugene H. Trinh

module with a connecting tunnel to *Columbia*'s crew compartment. The thirteen-day mission, replete with medical tests and materials processing experiments, was the first extended duration orbiter flight.

A close-up view of a monitoring screen in Spacelab shows a droplet of fluid adopting an almost perfectly spherical shape in microgravity.
▼

STS-46 *Atlantis*

Launched	July 31, 1992
Landed	August 8, 1992
Commander	Loren J. Shriver
Pilot	Andrew M. Allen

Mission specialists
Franklin R. Chang-Diaz, Jeffrey A. Hoffman,
Marsha S. Ivins, Claude Nicollier

Payload specialist
Franco Malerba

Milestones

This mission's main task was the deployment of
ESA's European Retrievable Carrier (EURECA) and
operation of the joint NASA-Italian Space Agency
Tethered Satellite System (TSS). EURECA was
boosted to its required higher orbit to be
retrieved by a future shuttle mission. The tether
experiment was supposed to sniff out
electrodynamic data from the space environment,
yet without the satellite component actually
becoming disconnected from *Atlantis*. The origins
of this project lay in the long-term fascination for
tethered systems expressed by many of the
world's space agencies. The great space futurist
Arthur C. Clarke championed vast orbiting
platforms tethered to the Earth itself, serving as
elevator shafts to orbit. Unfortunately, the tether
on this experiment failed to reel out properly.
After numerous attempts to free it, the satellite
was pulled back into the cargo bay and stowed for
return to Earth.

▲
*The tether experiment aboard STS-46 fails to deploy.
Similar experiments aboard a variety of space missions,
both Russian and American, have delivered, at best, mixed
results to date.*

*STS-47 crew member Jemison floats contentedly in the
Spacelab-J module during a rare moment of calm between
busy experimental schedules.*

▼

STS-47 *Endeavour*

Launched	September 12, 1992
Landed	September 20, 1992
Commander	Robert L. Gibson
Pilot	Curtis L. Brown Jr.

Mission specialists
Jay Apt, N. Jan Davis, Mae C. Jemison,
Mark C. Lee

Payload specialist Mamoru Mohri

Milestones

Spacelab-J was a joint venture between NASA
and the National Space Development Agency of
Japan (the forerunner of today's Japan Aerospace
Exploration Agency—JAXA). A Spacelab module
was used to conduct microgravity experiments in
materials and life sciences. The crew included the
first Japanese astronaut (Mohri) to fly aboard a
U.S. spacecraft, the first African-American woman
(Jemison) to fly in space, and the first married
couple (Davis and Lee) on the same mission and
in tne same vehicle.

An edged cutting tool from the Stone Age drifts inside the latest product of human ingenuity, in this philosophically stunning image from STS-52.

STS-52 *Columbia*

Launched	October 22, 1992
Landed	November 1, 1992
Commander	James D. Wetherbee
Pilot	Michael A. Baker

Mission specialists
Tamara E. Jernigan, William M. Shepherd, Charles L. Veach

Payload specialist
Steven G. MacLean

Milestones
This mission carried the U.S.-Italian Laser Geodynamic Satellite II (LAGEOS-II), a spherical, passive object studded on the outside with more than four hundred prisms that reflected ground-based laser beams back to their source from whichever direction they arrived. By clocking the few fractions of a second's time lag between transmission of the beam and reception of the reflected signal, monitoring stations on Earth measured the distance between the ground and the satellite with a margin of error not much greater than the length of a credit card.

STS-53 *Discovery*

Launched	December 2, 1992
Landed	December 9, 1992
Commander	David M. Walker
Pilot	Robert D. Cabana

Mission specialists

Guion S. Bluford Jr.,Michael R. Clifford,
James S. Voss

Milestones

The final shuttle mission dedicated to secret
Department of Defense payloads, plus some
unclassified secondary payloads, such as a hand-
held location targeting instrument and another run
for the CLOUDS instrumentation first flown on
Discovery during 1984.

*STS-54's Tracking & Data Relay Satellite is sent on its way
from the cargo bay of* **Endeavour.**

*STS-54 space walkers Runco (left) and Harbaugh get to grips
with new tools for working on the future space station.*

STS-54 *Endeavour*

Launched	January 13, 1993
Landed	January 19, 1993
Commander	John H. Casper
Pilot	Donald R. McMonagle

Mission specialists

Gregory J. Harbaugh, Susan J. Helms,
Mario Runco Jr.

Milestones

This mission deployed a fifth tracking and data
relay satellite, which was successfully placed in its
correct orbit by an IUS booster. *Endeavour*'s
payload bay carried the Diffuse X-ray
Spectrometer (DXS) to collect data on x-ray
radiation from sources in deep space. On day five
of the mission, Runco and Harbaugh conducted
an EVA to test foot restraints and other support
tools ahead of the great space station assembly
adventure soon to come.

Launched	April 8, 1993		**Launched**	April 26, 1993
Landed	April 17, 1993		**Landed**	May 6, 1993
Commander	Kenneth D. Cameron		**Commander**	Steven R. Nagel
Pilot	Stephen S. Oswald		**Pilot**	Terence T. Henricks

Mission specialists

Kenneth D. Cockrell, C. Michael Foale,
Ellen Ochoa

Mission specialists

Bernard A. Harris Jr., Charles J. Precourt,
Jerry L. Ross

Payload specialists

Hans Schlegel, Ulrich Walter

Milestones

The Atmospheric Laboratory for Applications and Science-2 (ATLAS-2) gathered data on the intimate relationship between the sun's energy output and the Earth's middle atmosphere, and especially the ozone layer: a major environmental concern two decades ago, before the current global warming crisis began to dominate the environmental agenda. A suite of instruments mounted on an exposed Spacelab pallet made the observations. The crew used the remote manipulator arm to deploy Shuttle Point Autonomous Research Tool for Astronomy-201 (SPARTAN-201), a free-flying science platform for studying the solar wind. As planned, it was drawn back into the payload bay after two days.

Milestones

This was the second Spacelab flight under German mission management. More than eighty experiments were conducted, covering materials and life sciences, technology applications, Earth observations, astronomy, and atmospheric physics. A German-built robotic arm manipulated some of the exposed hardware and tested the capture of free-flying objects, via remote control from Earth.

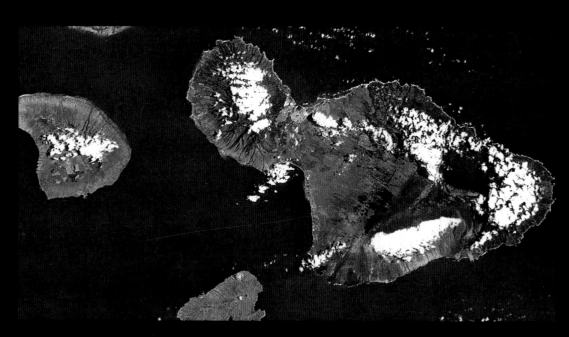

▲
STS-56's view of Maui, second-largest of the Hawaiian Islands, and a favorite vacation spot for astronauts.

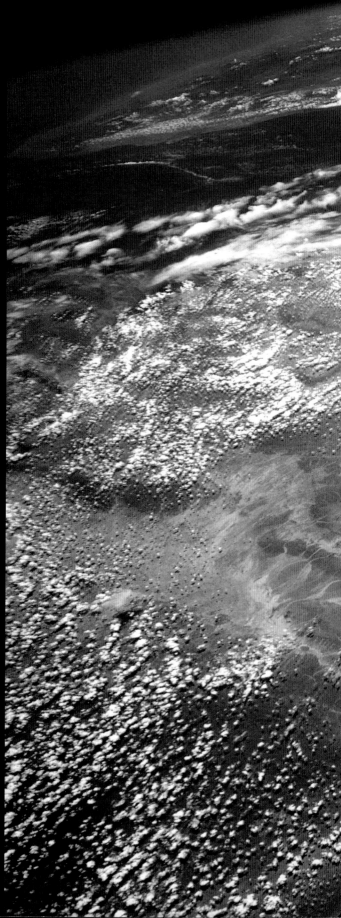

▶
The orange and tan colors of STS-55's oblique photo of the Horn of Africa show an arid landscape in the northern half of Somalia in east Africa. Most of the vegetation here is shrub brush and grasslands.

Commander   Ronald J. Grabe
Pilot   Brian Duffy

Payload commander
G. David Low

Mission specialists
Nancy J. Sherlock, Peter J. K. Wisoff,
Janice E. Voss

designed to more than double the available workspace for crew members. This highly adaptable system would be featured in many missions to come, and it presaged the modern age of increasing private development of space habitats and other live-aboard systems.

▶

Low and Wisoff simulate handling of large components in space. Low's feet are anchored to the robot arm, while Wisoff role-plays as the "component."

▼

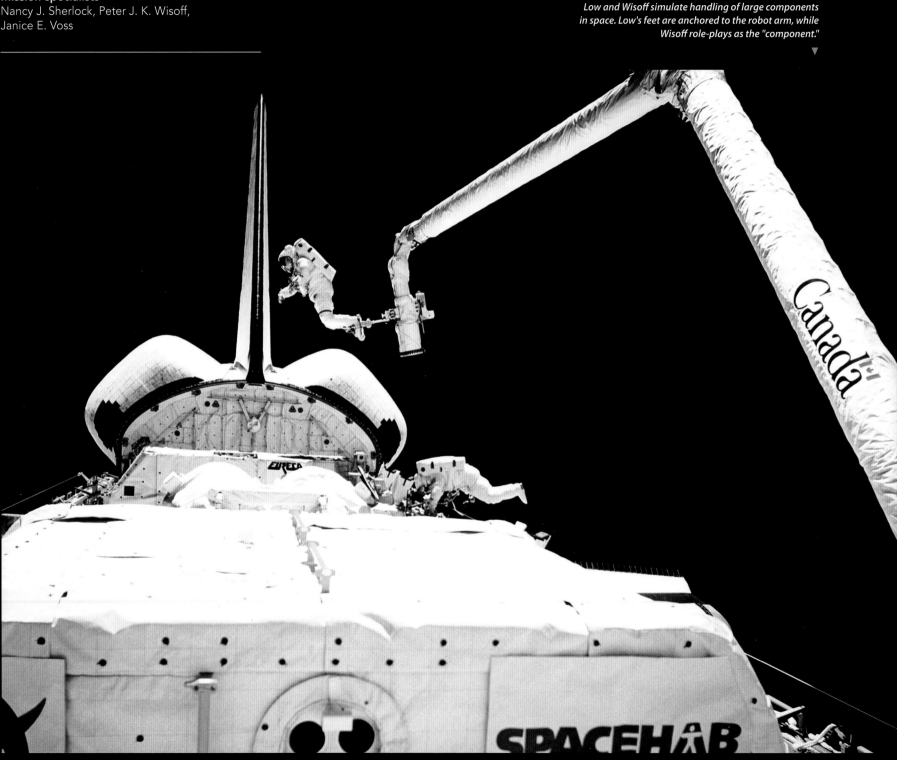

STS-51 *Discovery*

Launched	September 12, 1993
Landed	September 22, 1993
Commander	Frank L. Culbertson Jr.
Pilot	William F. Readdy

Mission specialists
Daniel W. Bursch, James H. Newman, Carl E. Walz

Milestones
The Advanced Communications Technology Satellite (ACTS) was deployed on the first day of flight. On day two, the crew deployed a second payload, the Orbiting and Retrievable Far and Extreme Ultraviolet Spectrograph-Shuttle Pallet Satellite (OERFEUS- SPAS), the first in a series of such missions. An IMAX film camera was mounted on the SPAS. During EVA activities, tools were evaluated for the impending first Hubble Space Telescope servicing mission. Here, Newman tests a foot restraint.

A drag chute slows STS-58 Columbia as it rolls to a perfect runway landing at the conclusion of the longest shuttle mission at that time.

STS-58 *Columbia*

Launched	October 18, 1993
Landed	November 1, 1993
Commander	John E. Blaha
Pilot	Richard A. Searfoss

Mission specialists
Shannon W. Lucid, William S. McArthur Jr.,

Milestones
Fourteen physiological experiments were conducted in Spacelab. Eight of the tests focused on the crew and the other six on forty-eight rodents. The crew collected hundreds of blood, urine, and other samples from themselves and the rodents, helping to build a picture of how humans and animals adapt to weightlessness.

STS-61 *Endeavour*

Launched	December 2, 1993
Landed	December 13, 1993
Commander	Richard O. Covey
Pilot	Kenneth D. Bowersox

Payload commander
F. Story Musgrave

Mission specialists
Tom Akers, Jeffrey A. Hoffman, Claude Nicollier,
Kathryn C. Thornton

Milestones

If any single shuttle mission can be said to have rebuilt NASA's damaged reputation in the wake of the *Challenger* disaster, this was it. During five long and complex space walks, two teams of astronauts repaired the blurry Hubble Space Telescope. Essentially, the machine was given a new set of internal optics that would, at last, bring the light collected by its flawed primary mirror into sharp focus on the sensor arrays. On the third day of the mission, Nicollier used the remote manipulator arm to grab Hubble and position it upright in *Endeavour*'s payload bay. On day four, Musgrave and Hoffman serviced some of Hubble's gyroscope guidance and attitude control equipment. The next day, Thornton and Akers replaced the solar arrays, discarding the old ones that had a tendency to buckle during the sudden temperature changes of sunrise or sunset.

The final two EVAs were devoted to the much-anticipated main business of the mission: the corrective optics systems, which had to be installed with perfect precision deep in the heart of Hubble. A few weeks later, the entire world celebrated as the telescope began its long and reliable career delivering fantastic views of deep space. This mission also demonstrated the benefits of human-aided construction for the forthcoming space station.

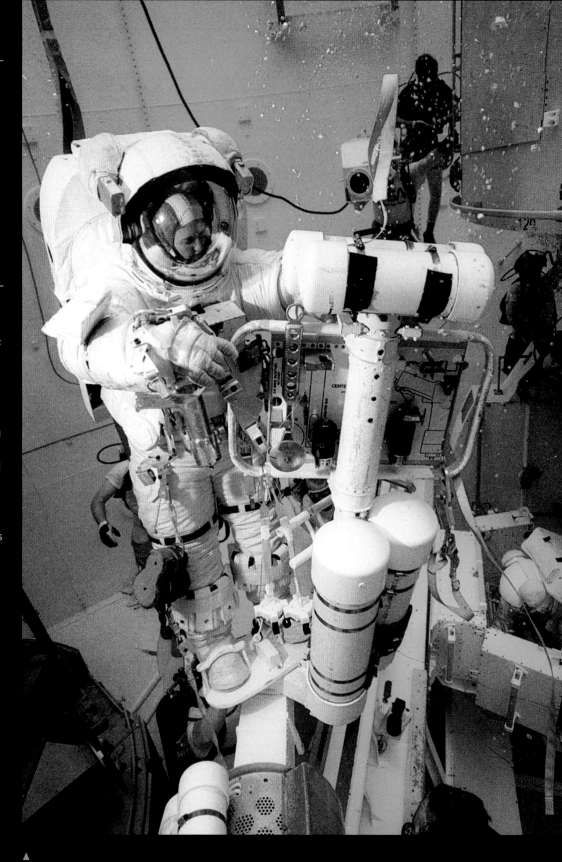

▲
Thornton trains in the Neutral Buoyancy water tank at the Johnson Space Center, prior to the historic STS-61 first Hubble Space Telescope servicing mission.

▶
This time performing her tasks for real, Thornton prepares equipment necessary to complete the Hubble servicing.

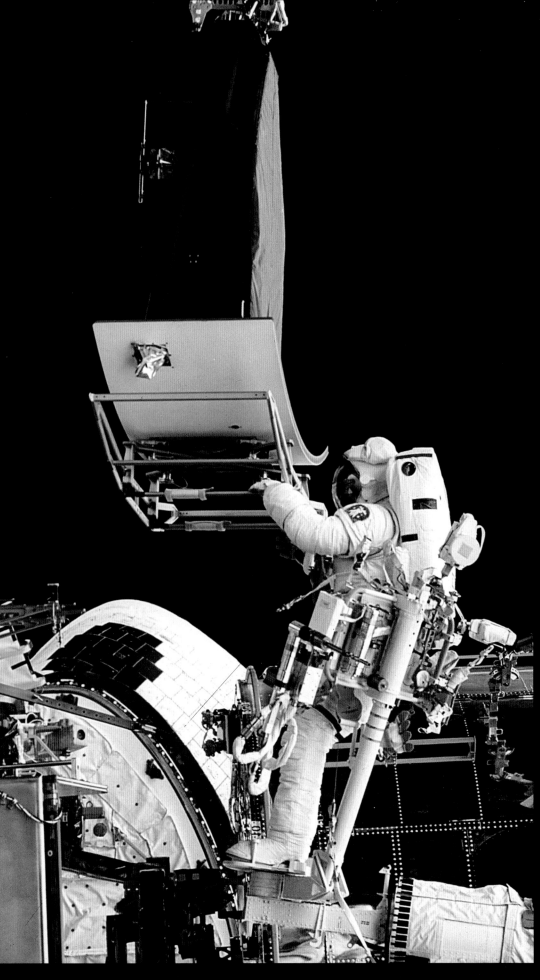

Hoffman holding the replacement equipment for Hubble's Wide Field/Planetary Camera-1, prior to the delicate operation of placing it inside the telescope.

Musgrave, anchored on the end of the robot arm, is lifted toward the top of Hubble, where he is about to install protective covers on the magnetometers.

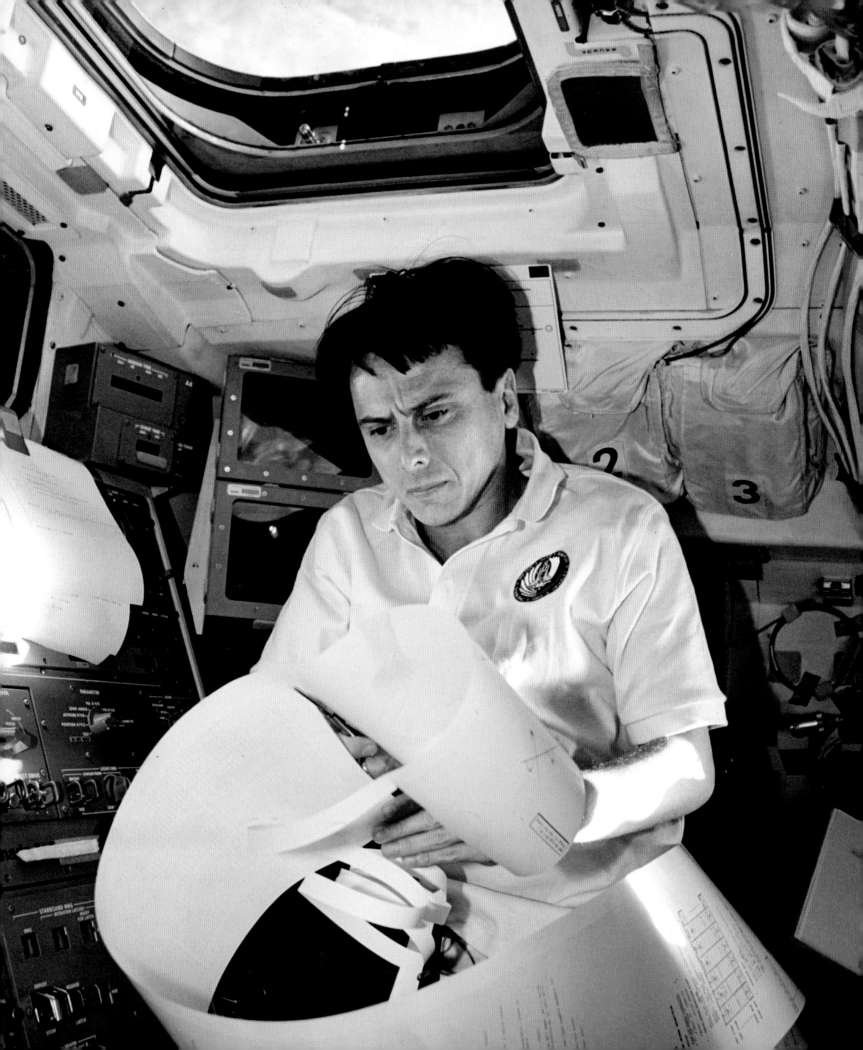

Even in space, there is no escaping the paperwork. On Discovery's aft flight deck, STS-60 astronaut Chang-Diaz sorts his way through one of the longest mail messages in shuttle history.

STS-60 *Discovery*

Launched	February 3, 1994
Landed	February 11, 1994
Commander	Charles F. Bolden Jr.
Pilot	Kenneth F. Reightler Jr.

Mission specialists
Franklin R. Chang-Diaz, N. Jan Davis, Ronald M. Sega, Sergei K. Krikalev

Milestones
For the first time in history, a Russian cosmonaut, Sergei Krikalev, flew into orbit aboard a U.S. spacecraft, as NASA began to implement a historic agreement on NASA and Russian Space Agency cooperation in human space flight: an accord that would culminate in shuttle visits to the Russian Mir space station and lead, eventually, to a long-term partnership for building the International Space Station. This mission also marked the second flight of the commercially developed SPACEHAB pressurized support module, loaded with scientific experiments.

STS-62 *Columbia*

Launched	March 4, 1994
Landed	March 18, 1994
Commander	John H. Casper
Pilot	Andrew M. Allen

Mission specialists
Charles D. Gemar, Marsha S. Ivins, Pierre J. Thuot

Milestones
Primary payloads for this mission were the U.S.-built Microgravity Payload-2 (USMP-2) and the Office of Aeronautics and Space Technology-2 (OAST-2) with a suite of six technology tests related to space station planning. The microgravity experiments included materials processing and crystal growth: a common theme in space research, as scientists and commercial entities alike seek new ways of exploiting microgravity for manufacturing and drugs development for industries back on Earth.

A rare low-light photograph of STS-62's flight deck with the lights switched off, as the crew prepare for re-entry and descent to landing. The instruments are illuminated.

STS-59 *Endeavour*

Launched	April 9, 1994
Landed	April 20, 1994
Commander	Sidney M. Gutierrez
Pilot	Kevin P. Chilton

Payload commander
Linda M. Godwin

Mission specialists
Jay Apt, Michael R. Clifford, Thomas D. Jones

Milestones

Endeavour's payload bay held a substantial suite of Earth atmospheric and ground terrain monitoring systems. The Space Radar Laboratory (SRL) included the space-borne imaging radar, a synthetic aperture radar, and an atmospheric instrument called Measurement of Air Pollution from Satellites (MAPS). The German and Italian space agencies built and operated the synthetic aperture sensors. The science communities of a dozen countries played some role in the mission.

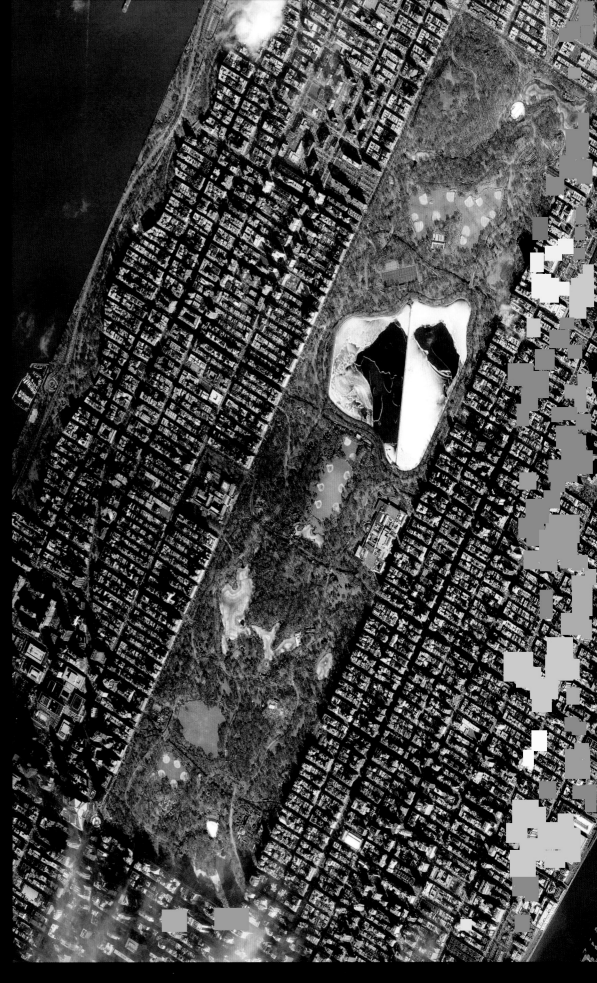

STS-59's detailed view of Central Park in New York shows the Jacqueline Kennedy Onassis Reservoir still frozen after a hard winter.

◀

Naito-Mukai floats through the hatch connecting Columbia's mid-deck with the Spacelab STS-65.

STS-65 *Columbia*

Launched	July 8, 1994
Landed	July 23, 1994
Commander	Robert D. Cabana
Pilot	James D. Halsell Jr.

Payload commander
Richard J. Hieb

Mission specialists
Leroy Chiao, Donald A. Thomas, Carl E. Walz

Payload specialist
Chiaki Naito-Mukai

Milestones
Mukai became the first Japanese woman to fly in space. More than eighty materials processing and life sciences experiments, representing two hundred scientists from six space agencies, were located in a Spacelab module. Fifty experiments delved into human physiology and the study of radiation's effects on biological tissues. The German Space Agency provided a centrifuge that investigated how organisms (jellyfish in particular) react to different gravity levels.

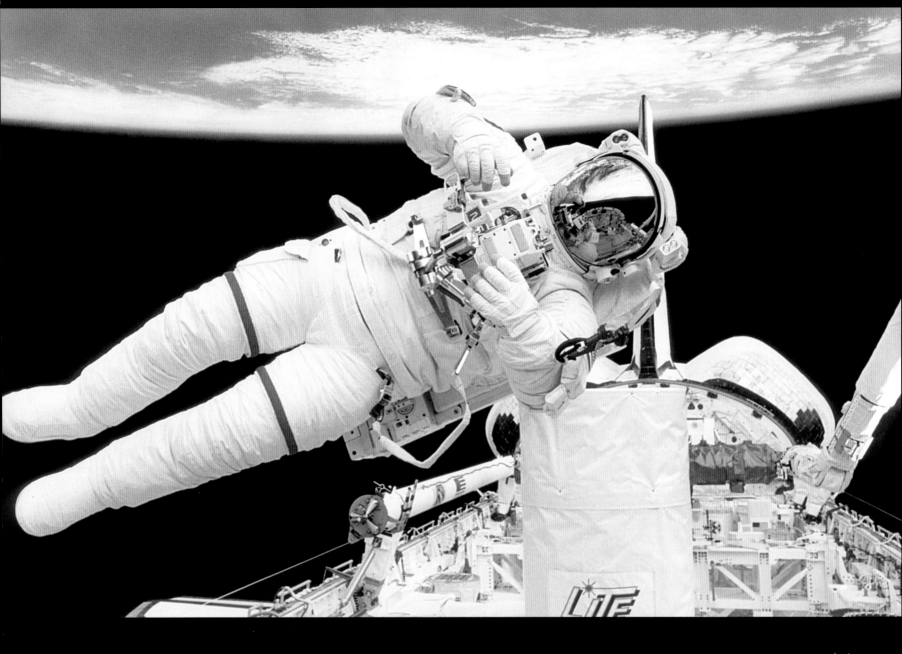

STS-64 *Discovery*

Launched   September 9, 1994
Landed   September 20, 1994
Commander   Richard N. Richards
Pilot   L. Blaine Hammond Jr.

Mission specialists

Susan J. Helms, Mark C. Lee, Jerry M. Linenger,
Carl J. Meade

Milestones

Discovery's main payload was the Light Detection and Ranging (LIDAR) system, a type of optical radar using laser pulses instead of radio waves to study Earth's atmosphere. Unprecedented views were obtained of storm systems, dust clouds, pollutants, and forest burning: a disturbing sight from space, especially in LIDAR's unforgiving scientific detail. An additional highlight was the first untethered space walk in a decade. Mission specialists Lee and Meade tested a new backpack, the Simplified Aid for EVA Rescue (SAFER),

designed for use in the event of a space-walking crew member accidentally going adrift from the normal system of tethers, foot restraints, and exterior handholds.

On day five, the Shuttle Pointed Autonomous Research Tool for Astronomy-201 (SPARTAN-201) free flyer was released using *Discovery*'s remote manipulator arm. Making its second flight, the SPARTAN collected data about the solar wind. The little platform was retrieved after two days.

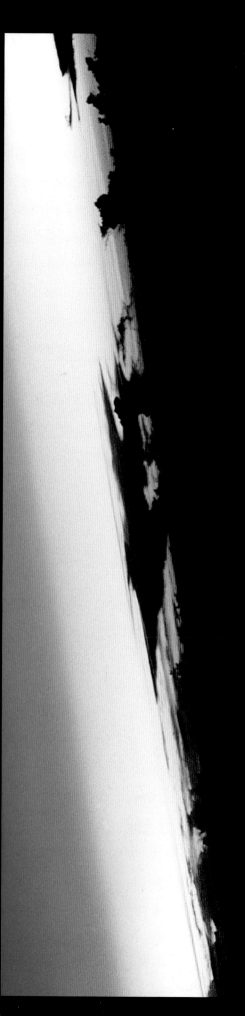

STS-68 *Endeavour*

Launched	September 30, 1994
Landed	October 11, 1994
Commander	Michael A. Baker
Pilot	Terrence W. Wilcutt

Payload commander
Thomas D. Jones

Mission specialists
Daniel W. Bursch, Steven L. Smith,
Peter J. K. Wisoff

Milestones

The second flight of the SRL, first flown on
Endeavour's STS-59 mission during April of that
same year, formed part of NASA's Mission to
Planet Earth. Flying SRL during different seasons
enabled changes in the Earth's climate between
the first and second flights to be observed. The
mission also studied fires deliberately set in British
Columbia and Canada for forest management.
The shuttle's maneuvering precision was
demonstrated when *Endeavour* was called upon
to make multiple passes over specific ground
targets, looking for infinitesimal alterations in
ground topography that might have occurred
since the last SRL flight. On one occasion,
Endeavour was piloted to within half its own
length of a specific location above the Earth that it
had passed through previously.

▶

*The long, dense plume of smoke and ash from the
Kamchatka peninsula's newly erupted volcano,
as osberved by the crew of STS-59.*

◀

*A telephoto shot of the Earrth's atmosphere edge-on,
showing clouds silhoutted against the dawn sunlight.*

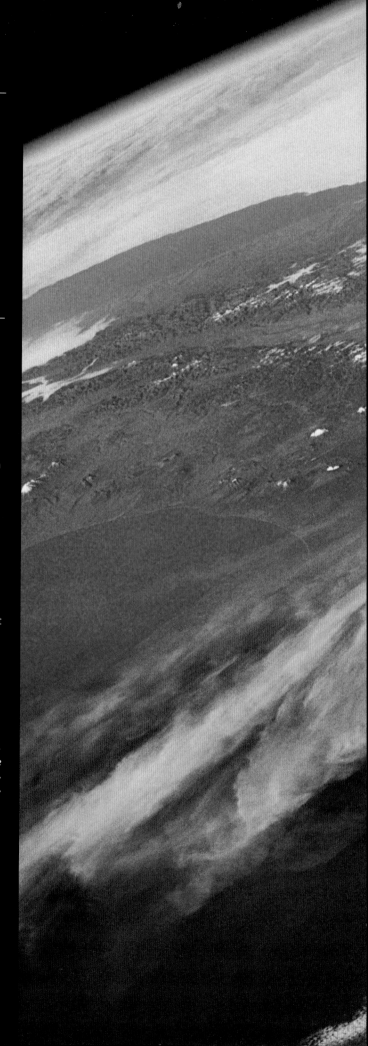

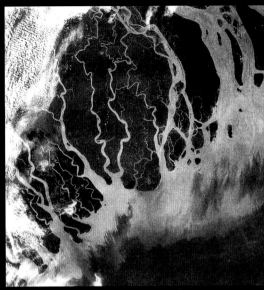

STS-66's view of the Ganges river delta, with its extensive mangrove mud flats, swamp vegetation, and sand dunes.

STS-66 *Atlantis*

Launched	November 3, 1994
Landed	November 14, 1994
Commander	Donald R. McMonagle
Pilot	Curtis L. Brown Jr.

Payload commander
Ellen Ochoa

Mission specialists
Jean-Francois Clervoy, Scott E. Parazynski,
Joseph R. Tanner

Milestones

This flight was a comprehensive effort to collect data about the sun's energy output, the chemical makeup of the Earth's middle atmosphere, and how these factors affect global ozone levels, using the Atmospheric Laboratory for Applications and Science (ATLAS) equipment mounted on exposed Spacelab pallets in the cargo bay. Another major payload was the Cryogenic Infrared Spectrometers and Telescopes for the Atmosphere-Shuttle Pallet Satellite (CRISTA-SPAS), a somewhat inelegant name for astronomical experiments operated jointly by NASA and the German Space Agency. CRISTA-SPAS took independent flight for eight days before being retrieved.

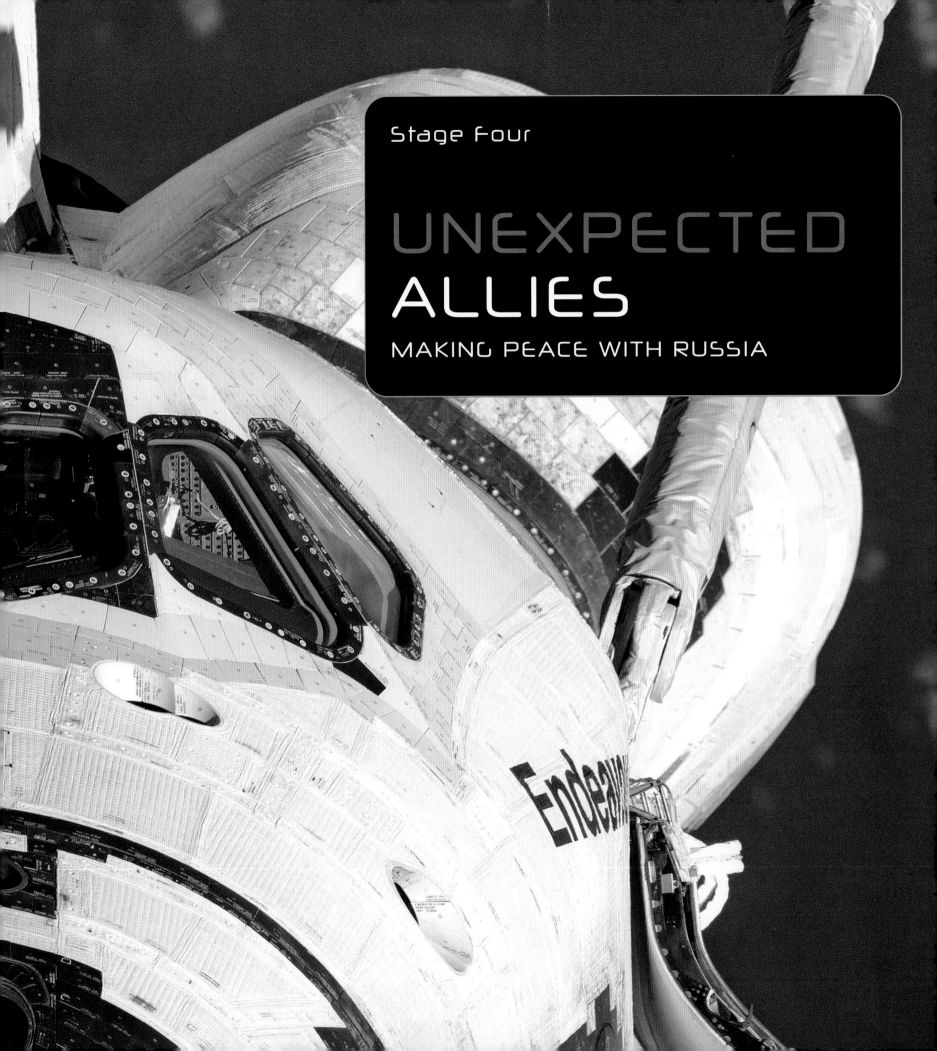

UNEXPECTED ALLIES

MAKING PEACE WITH RUSSIA

UNEXPECTED ALLIES

MAKING PEACE WITH RUSSIA

The great space adventures of the 1960s, from the first artificial satellite to the first men on the moon, were driven by fierce superpower competition. In the 1990s this rivalry was supplanted by collaboration.

THE ORBITAL REALM is as quiet and seemingly as unchanging as only a perfect and infinite vacuum can be. On Earth, amid all the air and noise, circumstances change fast. The Soviet empire fell away, the Cold War came to an end, and Reagan's Republican administration gave way to Bill Clinton's era. The fortunes of space chiefs also depend on the political tides. On June 6, 1992, the new NASA administrator Dan Goldin and his Russian opposite number, Uri Koptev, met for the first time in Goldin's Washington apartment. The subject under discussion was epic in its scope. Could Russia and America, those bitter Cold War enemies, find common cause in space now that the Soviet era had ended?

On February 19, 1986, Russia launched the first component of its orbiting Mir ("Peace") space station. On March 13, 1986, Mir's first crew, Leonid Kzim and Anatoly Solovyev, blasted off for a rendezvous and then clocked up an impressive roster of activities. After two months aboard Mir, they took their Soyuz capsule for a fifty-day visit to the old Salyut 7, which was still in orbit. They made two space walks, investigating the exterior condition of Salyut and practicing assembly work in space. Then they flew back to Mir, reoccupying the new station for three weeks before finally heading for home after a mission totaling 125 days. By any measure, the Soviets consolidated their superiority in Earth orbit using relatively inexpensive technology to achieve feats that NASA keenly wanted to match.

However, by 1992 and the time of that strange meeting between Goldin and Koptev, matters were very different. Post-Communist Russia was in social and economic turmoil, and the Mir station was operating not merely in the vacuum of space, but in an equivalent vacuum of financial and political resources. Russia was not willing to abandon its proudest space achievement, yet neither was it able to support this expensive project unaided. Turmoil on the ground was disrupting the launches of Soyuz crew capsules and Progress supply ferries for Mir. Goldin and the Clinton administration saw an opportunity to reach acorss the old Cold-War divide and find common ground. America, too, needed a helping hand: not with cash, of course, but with something just as valuable—a compelling political reason for its space station. Encouraging peace between superpowers seemed a very good rationale for it. But first, this new tone of collaboration had to be tested.

Rivals become colleagues

The meeting was not exactly secret, but neither was it widely heralded to the press. Such caution was understandable. At this early stage, neither man knew what he would find in the other. Despite barriers of nationality, history, and language, Koptev and Goldin quickly saw how similar their problems were. They were both ex–Cold War warriors who believed in exploring space for peaceful ends. They had both been stripped of much power by their political masters and forced into painful spending cuts. Dr. John Logsdon, a renowned space policy advisor in Washington, describes this meeting as "the two of them falling into administrator's love. They were almost mirror images of each other."

On February 3, 1994, cosmonaut Sergei Krikalev became the first Russian citizen to be launched aboard a U.S. spacecraft, shuttle *Discovery*. Exactly one year to the day later, *Discovery* lifted off to rendezvous with Russia's Mir, coming to within forty feet of the station, with astronaut James Wetherbee in command. Eileen Collins sat next to him on the flight deck, the first female astronaut to pilot a shuttle. As *Discovery* made its final approach, cosmonaut Vladimir Titov waved out of the flight deck window to his colleagues in Mir. A leaking thruster required NASA and the Russian Space Agency (RSA) to negotiate some last-minute revisions to the rendezvous. There were concerns that tiny amounts of fuel might contaminate Mir's delicate external instruments. The eventual solution to this problem set an encouraging precedent, as Russian and American crews and senior administrators alike adapted to circumstances, rather than

TRUSTY RUSSIAN WORKER

The Soyuz (Russian for "Union") was developed from a Soviet-era expendable intercontinental ballistic missile designed by the brilliant rocket pioneer Sergei Korolev in the late 1950s, and subsequently used to launch Sputnik, the world's first satellite (1957), and Yuri Gagarin, the first human space voyager (1961). In the late 1960s, the main stages of the rocket were adapted, and the upper stages lengthened, to take the weight of Russia's workhorse three-seater capsule, also known as Soyuz. This reliable hardware has kept Russia in the space business for more than half a century. Here we see what looks like a cluster of twenty engines, but this outer appearance masks an economical inner design. Fuel pumps feed four nozzles in each stage, maximizing thrust while minimizing the machinery behind the nozzles.

letting those circumstances stand in the way of completing a successful mission.

Five weeks later, on March 14, 1995, NASA astronaut Norman Thagard hitched a ride aboard a Russian Soyuz capsule and became the first American astronaut to board Mir. On June 29 shuttle *Atlantis* came to take him home, establishing a firm dock with Mir almost exactly twenty years since the (Nixon-approved) Apollo-Soyuz docking mission in the summer of 1975 had set a tentative precedent even as the Cold War was still going on. Two Russians on *Atlantis* relieved two others aboard Mir, who came home with Thagard aboard the shuttle. This was the first time in history that people had flown up into space aboard one type of craft and had come home in another.

These seismic shifts in history resulted in a substantial redesign of the as-yet unbuilt space station to include Russian modules and a new bias to its political justifications. The somewhat suggestive name Freedom was dropped, for there was no longer any suggestion that cash-strapped post-Soviet Russia might still be a rival in orbit. The project was renamed Alpha, and finally became known as the International Space Station (ISS). What had begun as an anti-Soviet foreign policy in the Reagan era assumed different purposes under President Clinton.

The geopolitics of space

The motivation for extending the hand of friendship to Russia was complicated. One factor was the need to protect its huge rocket industry from economic turbulence while a market economy slowly began to emerge, and most especially to prevent its expertise from reaching undesirable purchasers elsewhere in the world. On the other hand, collaborative arrangements promised that a wide range of excellent Russian launch vehicles would become available to the Western space community. The Clinton era also saw new commercial arrangements on home ground, between NASA and its major shuttle hardware and service suppliers. In 1995, a new commercial entity, the United Launch Alliance (ULA) was established as a joint venture between Boeing and Lockheed Martin, consolidating a dozen different areas of crew training and spacecraft launch preparations under one contractual banner. ULA became responsible for more than eight thousand shuttle-related personnel.

Space flight is often viewed by the public as a questionable expense, as if, somehow, all the dollars are squeezed into giant fireworks and burned. The truth is more complex. Space activities have far-reaching effects on the ground, in terms of jobs, industrial expertise, international prestige, and—perhaps most usefully of all—as a means of encouraging peace and healthy collaboration between powerful nations. American leadership has driven these positive developments, and the shuttle has been a central tool in that project. Now that it has gone, who will lead the clan of international spacefarers in the years and decades to come?

◄
A Soyuz rocket leaves the launch pad at Baikonur. The slender tower of the launch escape system at the tip of the stack shows that the payload on this occasion is a crewed version of the Soyuz capsule, rather than the robotic and uncrewed "Progress" cargo variant.

▶
The crew compartment of a Soyuz spacecraft accommodates three people at most. Sometimes, one crew couch is sacrificed to make way for supplies and equipment.

A Soyuz crew transfer vehicle heads toward docking. At front, and tipped by the docking probe, is the egg-shaped orbital module that provides a little extra working room for the crew. The middle section is the re-entry capsule, and the white rear module contains propulsion systems.

STS-63 *Discovery*

Launched   February 3, 1995
Landed   February 11, 1995
Commander   James D. Wetherbee
Pilot   Eileen M. Collins

Mission specialists
C. Michael Foale, Bernard A. Harris Jr.,
Vladimir G. Titov, Janice E. Voss

Milestones
This was the first flight of a female shuttle pilot (Collins) and, as part of the emerging International Space Station program, the second flight to orbit with a Russian cosmonaut on board the shuttle. The highlight was the first close approach and cautious flyaround of Russia's space station Mir, testing the ability of the two nations eventually to dock those two craft together.

Atlantis experienced minor propellant leaks from its thruster maneuvering system, but the crew dealt with the problem by reassigning valves and switching to backup thruster procedures. After extensive negotiations between NASA and the Russian space team, permission was granted for close approach. Titov, who had already spent a long tour of duty aboard Mir (more than a year), communicated happily with three cosmonauts aboard the station: Commander Alexander Viktorenko, Flight Engineer Elena Kondakova, and physician Valery Polyakov.

As Wetherbee nudged *Discovery* up close to Mir, he radioed across to the other spacecraft: "We are bringing our nations closer together. The next time we approach, we will shake your hand and together we will lead our world into the next millennium."

"We are one. We are all humans," Viktorenko responded from within the Mir. Toward the end of the mission, Foale and Harris began a simple but necessary EVA, testing modifications to their spacesuits designed to keep space walkers warmer in the extreme cold of space. The results were mixed, and during a pass around the night side of the Earth, both astronauts reported an uncomfortable chill. Harris at least had the satisfaction of becoming the first African-American to walk in space.

▲
STS-63's view of the Mir platform. Sometimes derided by the Western press as outdated, Mir achieved many technological milestones long before the first modules of the International Space Station were even built, let lone launched to orbit.

Valeriy Polyakov, who boarded Mir in January 1994, looks out the window during the rendezvous.
▼

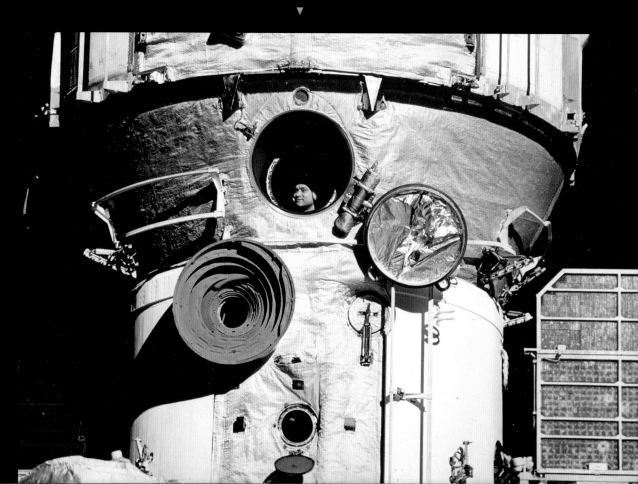

STS-67's ASTRO instruments streamed data to Earth, and distributed it around the world via the Internet.

▼

STS-67 *Endeavour*

Launched	March 2, 1995
Landed	March 18, 1995
Commander	Stephen S. Oswald
Pilot	William G. Gregory

Payload commander
Tamara E. Jernigan

Mission specialists
John M. Grunsfeld,
Wendy B. Lawrence

Payload specialists
Samuel T. Durrance, Ronald A. Parise

Milestones
Endeavour completed the longest shuttle flight to date, allowing sustained examination of the universe seen in ultraviolet light, using the ASTRO suite of astronomical instruments as flown on STS-35 in December 1990. Three ultraviolet telescopes were mounted on a Spacelab pallet in the cargo bay. This was the first shuttle mission overtly connected to the Internet. Users of more than two hundred thousand computers around the world logged on to the ASTRO homepage at NASA's Marshall Space Flight Center. More than 2.4 million hits were recorded.

STS-71 *Atlantis*

Launched	June 27, 1995
Landed	July 7, 1995
Commander	Robert L. Gibson
Pilot	Charles J. Precourt

Mission specialists
Ellen S. Baker, Bonnie J. Dunbar,
Gregory J. Harbaugh

Mir crew to orbit
Anatoly Solovyev, Nikolai Budarin

Mir crew to Earth
Norman E. Thagard, Vladimir Dezhurov,
Gannady Strekalov

Milestones
STS-71 marked a number of historic firsts in spaceflight. It was NASA's 100th human space launch and the first secure docking with Mir, briefly creating the largest combined spacecraft ever in orbit. This mission also saw the first switch of shuttle crew members during the orbital phase. Homeward-bound Mir personnel included NASA astronaut Thagard, the first U.S. citizen to be carried into space on a Russian rocket and the first to conduct a long-duration mission aboard a Russian spacecraft. Just before *Atlantis* undocked, the cosmonauts still on board Mir temporarily abandoned their station, flying away from it in their Soyuz spacecraft to record unprecedented images of *Atlantis* and Mir separating.

STS-71 commander Gibson (foreground) shakes the hand of cosmonaut Vladimir Dezhurov, Mir's commander at the time of this historic first Shuttle-Mir docking.

▼

Inside the Spacelab module aboard the shuttle, the crewmembers of STS-71, Mir-18, and incoming crew for the Mir-19 mission, pose for an inflight picture. The socked feet of Anatoly Solovyev are at bottom center. Clockwise from him are Harbaugh, Gibson, Precourt, Budarin, Baker, Dunbar, Thagard, Strekalov, and Dezhurov.

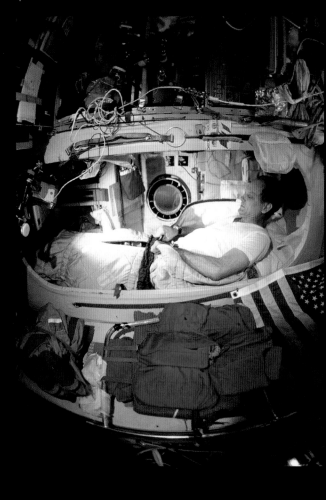

Thagard in his borrowed sleep station aboard Mir, looking forward to homecoming at the end of his long stay.

One of the most iconic shuttle pictures on record: Atlantis docked to Mir, with the Earth below rounded into an almost complete globe by virtue of a wide-angle lens.

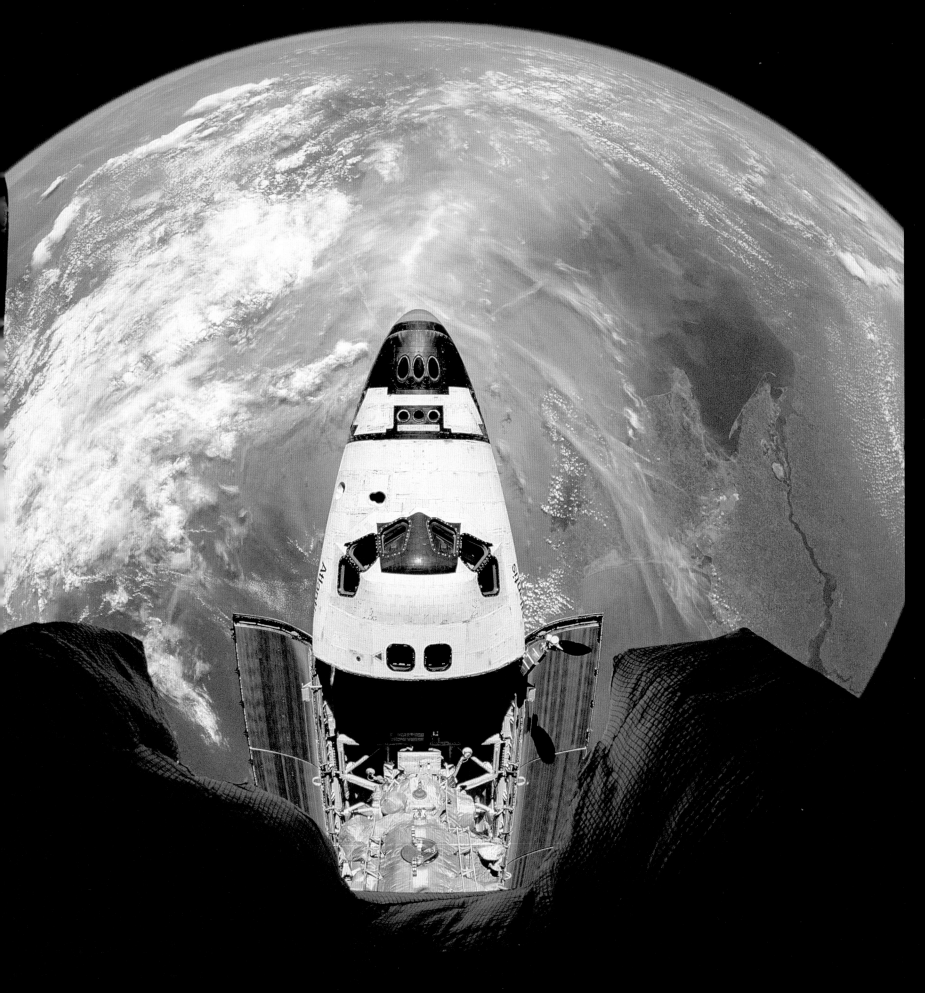

An unprecedented picture, taken from the Russian Soyuz
crew capsule that has temporarily separated from Mir
specifically to observe the moment of undocking
between the station and the shuttle.

STS-70 *Discovery*

Launched	July 13, 1995
Landed	July 22, 1995
Commander	Terence T. Henricks
Pilot	Kevin R. Kregel

Mission specialists
Nancy Jane Currie, Donald A. Thomas,
Mary Ellen Weber

Milestones

A tracking and data relay satellite was deployed from *Discovery*'s payload bay about six hours after liftoff. Approximately one hour later, its booster completed the first of two scheduled rocket burns to place the satellite in geosynchronous orbit as an operational backup, completing an existing TDRS network.

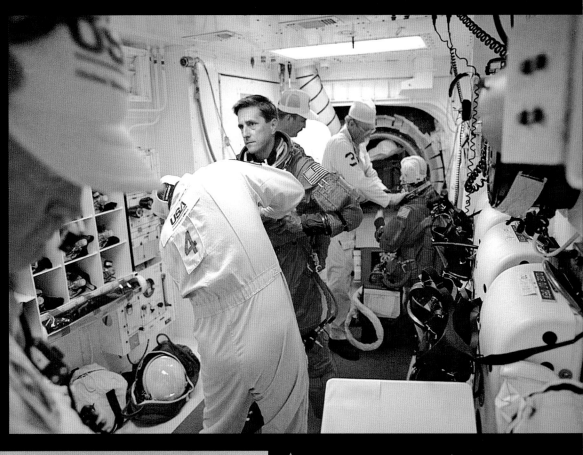

▲
After months of training, this is the real thing, as pad technicians give STS-70 Mission Specialist Thomas a final check before he enters the shuttle for takeoff.

◄
It's all smiles for Henricks and his STS-70 crew at the end of a successful mission capped by a smooth touchdown.

The Wake Shield is deployed to exploit the extreme vacuum for semiconductor rsearch.

STS-69 *Endeavour*

Launched	September 7, 1995
Landed	September 18, 1995
Commander	David M. Walker
Pilot	Kenneth D. Cockrell

Payload commander
James S. Voss

Mission specialists
Michael L. Gernhardt, James H. Newman

Milestones
Two different payloads were deployed and recaptured: a SPARTAN solar physics free-flier and a wake shield experiment, a twelve-feet-diameter stainless-steel disk designed to generate an extreme vacuum environment, literally in its wake, within which to grow thin and very pure semiconductor films for advanced electronics. On the tenth day of the flight, Voss and Gernhardt conducted a six-hour space walk to evaluate improvements in their spacesuits.

STS-69 photographs Hurricane Marilyn as it moves over
Puerto Rico. Another hurricane, named Luis, followed
a similar path earlier in the flight.

Astronaut Voss on EVA, attached to the ever-reliable
robot arm, and holding a camera.

STS-73 *Columbia*

Launched   October 20, 1995
Landed   November 5, 1995
Commander   Kenneth D. Bowersox
Pilot   Kent V. Rominger

Payload commander:
Kathryn C. Thornton

Mission specialists
Catherine G. Coleman, Michael E. Lopez-Alegria

Payload specialists
Fred W. Leslie, Albert Sacco Jr.

Milestones
The second flight of the U.S. microgravity laboratory, built on the success of its predecessor, flown on *Columbia* during Mission STS-50 in 1992. Research was conducted in fluid physics, materials science, biotechnology, combustion science, and commercial space processing.

These STS-73 images show a change, over time, in the shape of the interface between a liquid and a gas in a sealed container. Under the force of Earth's gravity, the interface would remain nearly flat, but in microgravity, the interface shape changes significantly. Major shifts of the liquid arise from small asymmetries in the container's shape.

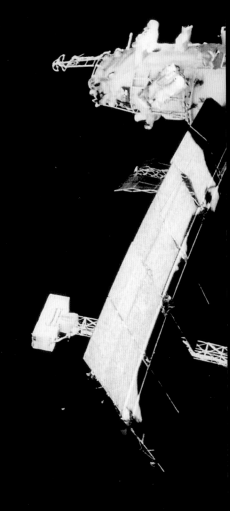

Working in support of the U.S. Microgravity Laboratory mission are Coleman (at the glovebox) and Leslie.

AN ISLAND
IN THE SKY

ESTABLISHING THE SPACE STATION

AN ISLAND IN THE SKY
ESTABLISHING THE SPACE STATION

The world's greatest and most complex peacetime space structure is even now flying overhead, and not a single day passes without the presence of humans in orbit. The "Space Age" did not end with the passing of Apollo.

In May 1983 the German minister for research and technology, Heinz Riesenhuber, visited Washington; met with NASA's latest administrator, James Beggs; and discussed with him the prospects of European collaboration in a space station. In March 1984 the Italian Prime Minister, Bettino Craxi, reinforced that message, personally advising President Ronald Reagan that Italy was keen to help build a space station and expand on the experience gained by European industry during the construction of Spacelab. On January 25, 1984, Reagan announced that a space station should be built—not by the United States alone, but by a partnership of allies. "Tonight, I am directing NASA to develop a permanently crewed space station, and to do it within a decade. We want our friends to help us meet this challenge and share in the benefits. NASA will invite other countries to participate so we can strengthen peace, build prosperity, and expand freedom for all who share our goals."

Almost exactly a year later, the ESA Council of Ministers met in Rome to discuss (among other matters) the station, then known as Freedom. The council was in favor of participating, provided that the international nature of the project was stressed. An official statement was published, which stated: "Europe will seek appropriate participation, with access and use of all elements of the station on a non-discriminatory basis." But first, NASA would have to earn the hoped-for support from its potential partners abroad; and those partners would expect something in return. The strength of European feeling also reflected a fear of America's traditionally unpredictable attitude toward long-term space projects. However, back in 1984, very

few people could have guessed how the station would be drastically reshaped, even at the planning stage, by seismic events on the ground. After 1989 and the rapid collapse of the Soviet empire, the consequent shifts in international relationships resulted in a redesign of the as-yet unbuilt space station to include Russian modules, accompanied by a new bias to its political justifications. The somewhat suggestive Reaganite name "Freedom" was dropped, for there was no longer any suggestion that cash-strapped post–Soviet Russia might still be a rival in orbit. The project was renamed Alpha, and finally became known as the International Space Station (ISS).

THE GRANDEST PROJECT

The International Space Station (ISS) is the largest engineering project ever conceived. Built by the U.S., Russia, Japan, Canada, and eleven European countries, the Station is powered by solar arrays generating 110 kilowatts. Its modules support a crew of at least six astronauts. On-board research is aimed at producing new materials and pharmaceutical products that can only be developed in microgravity. Assembly of the giant structure called for transportation of materials by the shuttle and Soyuz crew carriers, Progress cargo ferries, Proton heavy-lift rockets, European Ariane rockets with Automated Transfer Vehicles, and similar Japanese-built robot supply ships.

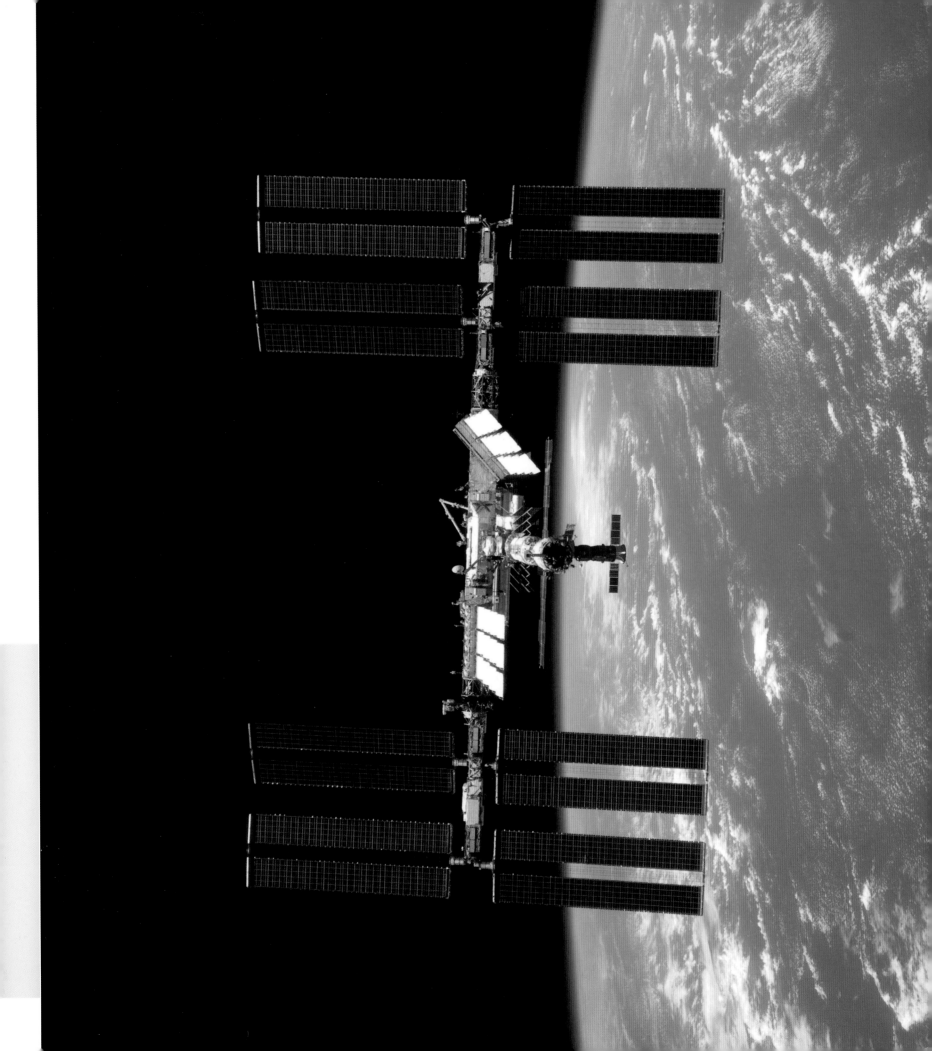

The first component of the ISS, the Russian-built control module *Zarya*, was sent into orbit by a Proton rocket launched from Baikonur in Kazakhstan, on November 20, 1998. Two weeks later, the shuttle, with its capacious payload bay and heavy-lift capacity, at last began to reveal its finest colors, as a true workhorse delivering further modules for the ever-expanding new space station, and rotating its crew members.

Colonizing space

The phrase "space station" conjures up an image of a metal cabin more or less the size of a three-storey house, manned by a select group of technicians, or perhaps one of the great Wernher von Braun's giant rotating wheels, hundreds of feet in diameter, with several dozen crew aboard. But some notable theorists have imagined gigantic stations bigger than many towns on Earth, serving social rather than scientific ends. Huge populations might live their entire lives far above the Earth, or so the theory goes.

In 1903 the great Russian pioneer of astronautics, Konstantin Tsiolkovsky (who worked on detailed ideas about rockets from the 1880s until his death in 1935), proposed a huge habitable cylinder, spinning on its axis and containing a greenhouse with a self-supporting ecological system; and in his novel *The World, the Flesh and the Devil* (1929), J.D. Bernal devised Bernal Spheres, self-supporting "worldships" capable of housing many thousands of inhabitants.

These colossal concepts were revived in the 1970s by an American academic, Gerard K. O'Neill at Princeton University. In the wake of the Apollo lunar landing of 1969, O'Neill proposed huge colonies in space as a way of easing the population pressures on Earth. The structures would be perhaps three kilometers in length, fabricated from materials processed in lunar factories. At first, O'Neill's ideas were intended primarily as a theoretical exercise to stretch the imaginations of his physics students, but the space colony theme hit a nerve during the 1970s with its detailed studies of closed ecological systems: an urgent concern for environmental campaigners. Especially popular was O'Neill's proposal that all heavy industry and energy production should be taken into space, thus saving the Earth from the burdens of pollution.

Smaller versions of O'Neill's space world might soon be built as tourist venues. It would require only a few hundred rich holidaymakers per year to make a small orbiting hotel worthwhile, provided that a suitable passenger-carrying launch vehicle was already available and safe to fly. Development of a genuinely reusable and fuel-efficient space transport system remains the last great unconquered space frontier. When flying into space is no more expensive or dangerous than stepping aboard a Boeing 747 for a trip across the Atlantic, then that's the moment when the future starts.

Making use of weightlessness

Centrifugal rotation has been a favorite theme for many space station theorists, from Tsiolkovsky in 1903 to O'Neill seventy years later. The spin would produce an artificial gravity. The current ideal is to exploit rather than counteract the weightless environment in order to conduct valuable scientific procedures that are impossible on Earth. Where once the gentle spinning of a space station was seen as a way of making life more or less bearable for its crew, total weightlessness is now seen as the main justification for building one in the first place. This is the ultimate explanation for why Skylab (NASA's first space station in the mid-1970s), Mir, and now the International Space Station look nothing like the speculative visions of the early space theorists.

Some of the old dreams do seem to be coming true, however. Those people who are frustrated by the apparently slow development of our space effort since the glory days of Apollo might take comfort from an interesting and simple statistic: today, more people go into space than in the 1960s, at the supposed "height" of the Space Age.

STS-88 approaches Zarya, with the new Unity module positioned upright on the end of the robot arm and poised for docking.

STS-88 *Endeavour*

Launched	December 4, 1998
Landed	December 15, 1998
Commander	Robert D. Cabana
Pilot	Frederick W. Sturckow

Mission specialists
Nancy J. Currie, Sergei K. Krikalev,
James H. Newman, Jerry L. Ross

Milestones

This was the shuttle's first International Space Station (ISS) flight, a twelve-day mission to begin assembly of the sprawling orbital platform. NASA's thirteen-ton Unity module was first connected to *Endeavour*'s docking collar. Then, using the robot arm, Russia's Zarya control module (the first ISS component, launched earlier aboard a Russian Proton rocket) was captured from orbit and mated to Unity, after some juggling to persuade the connectors to activate properly. Astronauts Ross and Newman conducted three space walks to attach power cables and handrails, while Krikalev, who was closely familiar with the Russian hardware, monitored events from within *Endeavour*'s cabin.

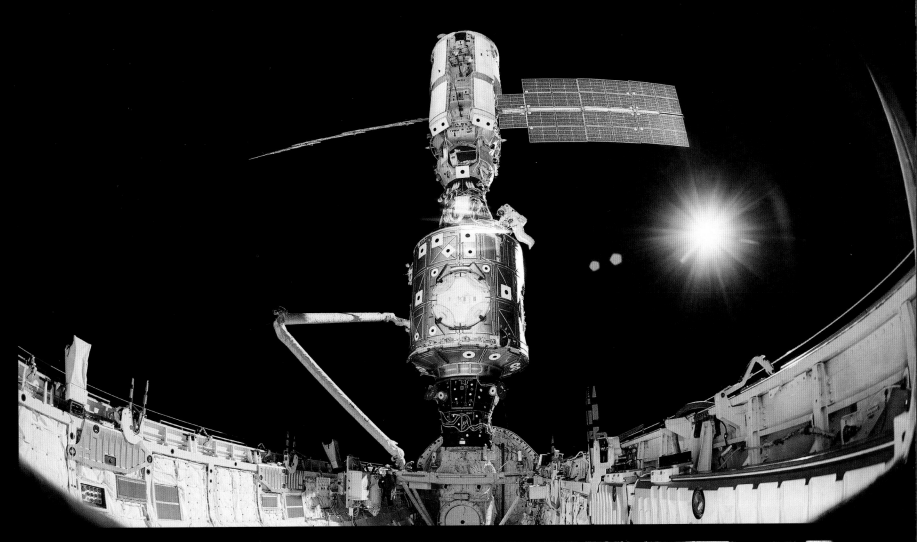

At left and above, wide angle before and after shots of the docking with Zarya to create the first multi-module structure for the Space Station.

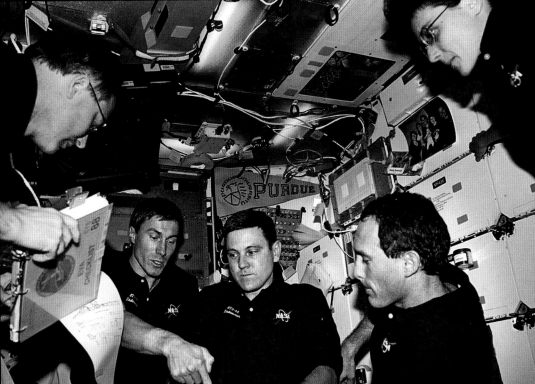

A planning session on Endeavour during STS-88. From the left are astronaut Ross, cosmonaut Krikalev, and astronauts Cabana, Newman, and Currie. Missing from the scene is Sturckow, who took the photo.

The nascent Space Station is released back into orbit. At the top of this view, the Unity module's mating adapters can be seen, with four vacancies awaiting modules that will be "berthed" to them on later delivery missions.

Three STS-88 crew members in one of the Pressurized Mating Adapters connected to the Unity and Zarya modules. In the foreground is Ross, with Cabana at middle left, and Krikalev on the right.

Newman, holding on to a handrail, waves at the camera during the first of three EVAs performed during the mission.

STS-96 *Discovery*

Launched   May 27, 1999
Landed   June 6, 1999
Commander   Kent V. Rominger
Pilot   Rick D. Husband

Mission specialists
Daniel T. Barry, Tamara E. Jernigan, Ellen Ochoa, Julie Payette, Valery Ivanovich Tokarev

Milestones
On May 29, *Discovery* made the first hard-docking to the nascent space station. Rominger made a textbook linkup with Unity's pressurized mating adapter as *Discovery* and the station flew over the Russian-Kazakh border. Jernigan and Barry subsequently undertook an EVA to install crane-like devices to the outside of the station, along with new portable foot restraints to fit both American and Russian space boots. After *Discovery*'s undocking, and a flyaround inspection of the station, mission specialist Payette deployed the STARSHINE satellite from the cargo bay. The small probe instantly became visible from Earth as part of a project allowing more than twenty-five thousand students from eighteen countries to track its progress.

▶
This spectacular photo shows the liftoff of Discovery STS-96, the second International Space Station assembly and resupply flight, and the first to dock with the new outpost.

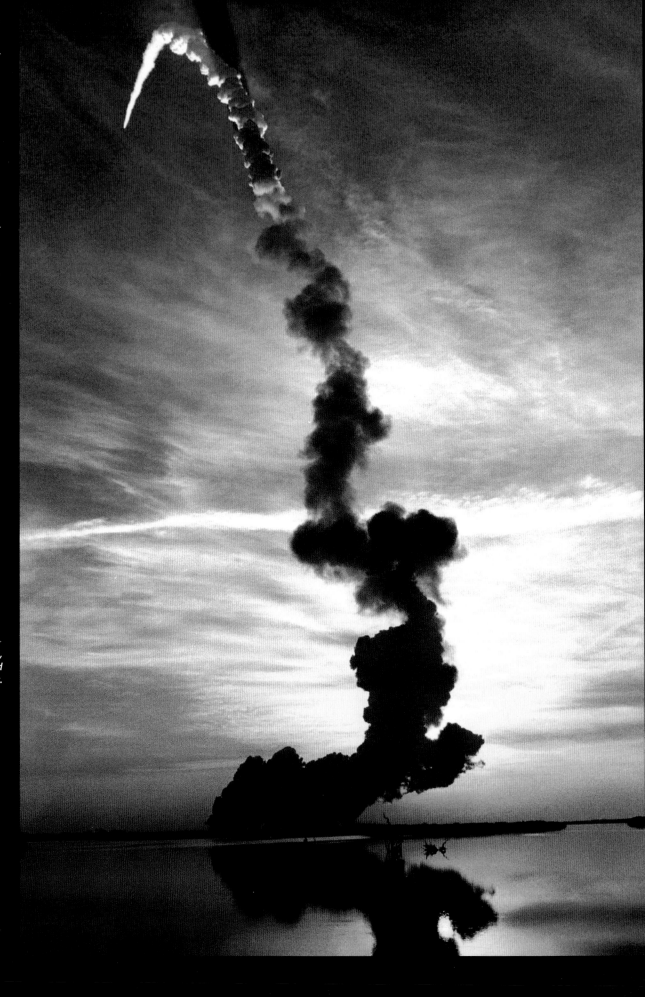

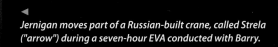

Jernigan moves part of a Russian-built crane, called Strela ("arrow") during a seven-hour EVA conducted with Barry.

The young Space Station recorded with a 70mm camera by STS-96 after undocking and separation of the two craft. Russia's Lake Baykal, not far from the Mongolian border, can be seen in the horizon landscape of the Earth.

Jernigan adjusts part of a Russian-built crane, called Strela ("arrow"). Her feet are anchored on a mobile restraint connected to the Remote Manipulator System.

Launched	July 23, 1999
Landed	July 27, 1999
Commander	Eileen M. Collins
Pilot	Jeffrey S. Ashby

Mission specialists
Catherine G. Coleman, Steven A. Hawley,
Michel Tognini

This mission, the first to be commanded by a woman (Collins), successfully deployed the Chandra x-ray observatory into its desired orbit, boosted by its two-section IUS. A secondary payload on board *Columbia*, the Southwest Ultraviolet Imaging System (SWUIS), captured ultraviolet imagery of the Earth, the Moon, Mercury, Venus, and Jupiter, while a high-definition TV camera was tested for future use on both the shuttle and the Space Station.

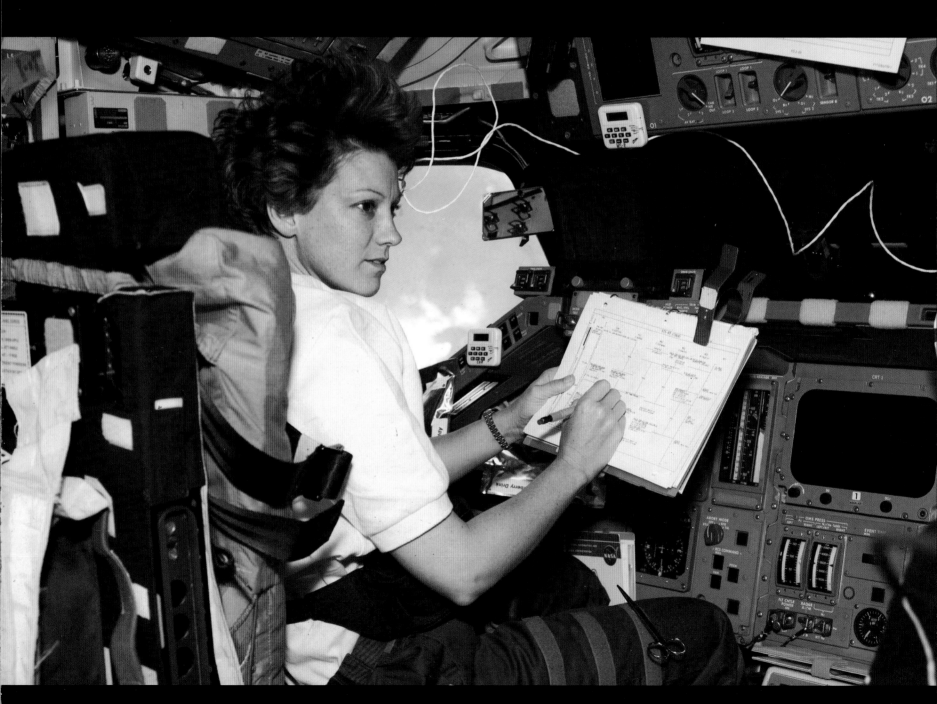

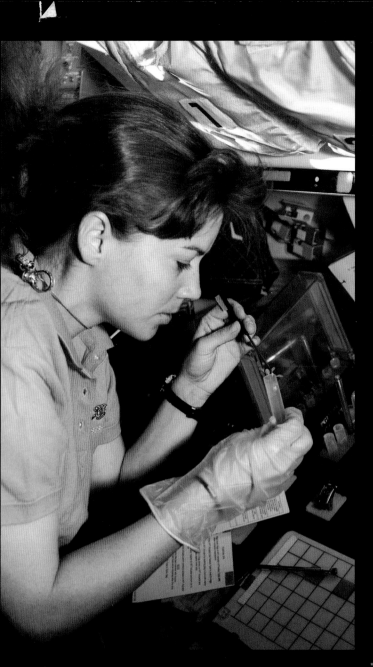

Coleman works with a small plant seedling on Columbia's flight deck, during one of many experiments conducted during the shuttle era to try and determine the effect of microgravity on the growth and behavior of living things.

Collins makes notes in her log at her station on STS-93's flight deck. The left-hand seat of any multi-person aircraft or spacecraft is almost always reserved for the captain or mission commander.

This image of Columbia's main engines, taken shortly after touchdown at the end of the STS-93 mission, gives a good view of these huge propulsion bells. However, the shot was taken for serious reasons. Engineers needed to check what appeared to be three small holes inside one of the bells.

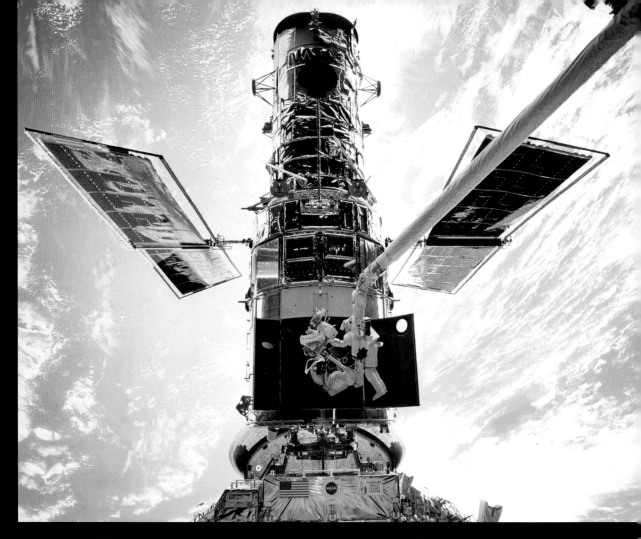

▶

Smith and Grunsfeld appear as small figures in this wide scene photographed during an EVA to replace gyroscopes inside Hubble.

STS-103 *Discovery*

Launched	December 19, 1999
Landed	December 27, 1999
Commander	Curtis L. Brown
Pilot	Scott J. Kelly

Mission specialists
Jean-Francois Clervoy, C. Michael Foale,
John M. Grunsfeld, Claude Nicollier,
Steven L. Smith

Milestones

The third Hubble Space Telescope servicing mission upgraded its systems, preparing the decade-old observatory for yet another decade of astronomical observations. After a complex thirty-orbit chase, Brown and Kelly maneuvered *Discovery* to a point directly beneath Hubble, then moved upward toward it. Clervoy grappled Hubble using the robotic arm and placed it in the rear of the cargo bay. Smith and Grunsfeld conducted the first EVA, replacing the telescope's attitude sensor gyroscopes and installing better bridges between Hubble's solar panels and its six ten-year-old batteries. During the mission's second space walk, Foale and Nicollier installed a new advanced computer. Half an hour later later, Hubble began "thinking" with its modernized brain. A final EVA by Smith and Grunsfeld was dedicated to installing a new radio transmitter. Hubble was released on Christmas Day.

▶

European Space Agency astronaut Clervoy controls STS-103 Discovery's robot arm during operations with the Hubble Space Telescope.

Reflections in a golden eye: Discovery's cargo bay, and the Earth's horizon, are reflected in Grunsfeld's helmet visor.

Smith retrieves a power tool while standing on the foot restraint. EVA photos are usually taken with wide angle lenses, but this shot is a telephoto view with a narrow depth of focus.

169

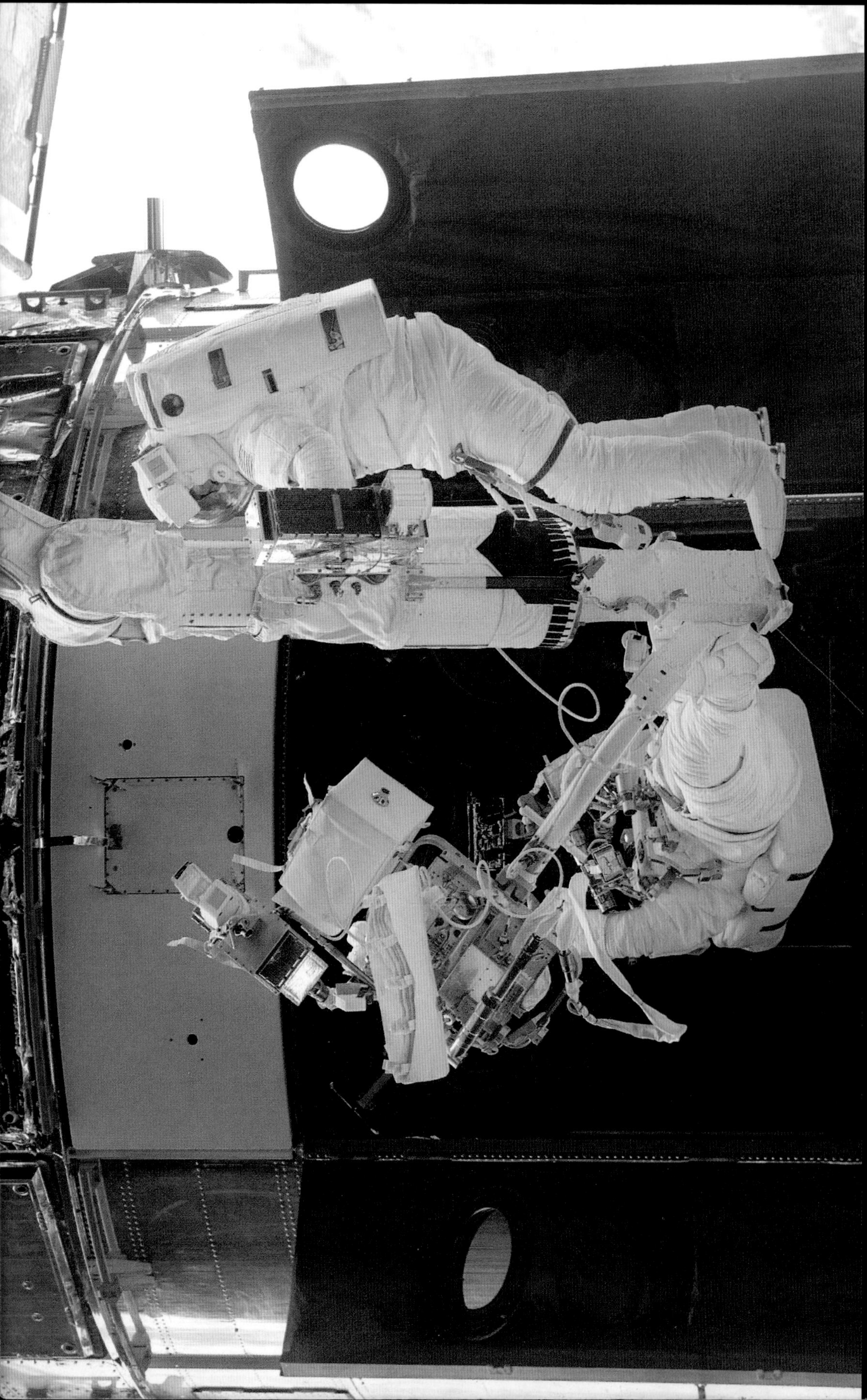

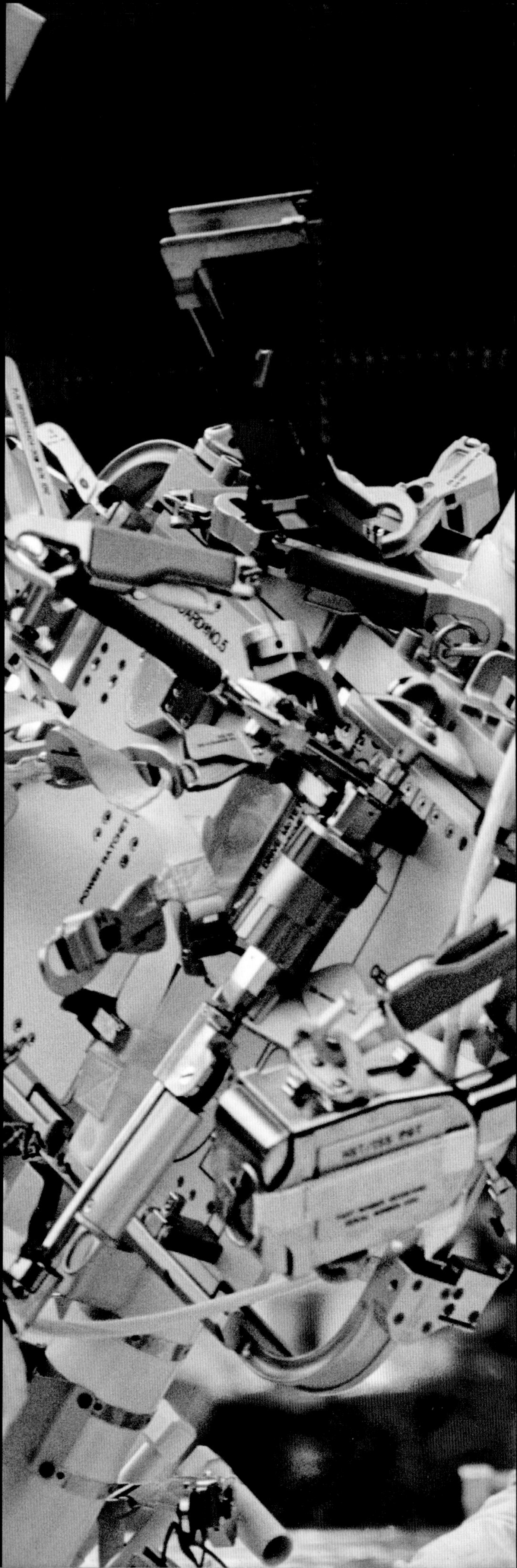

STS-103 space walkers Smith (right) and Grunsfeld work cautiously around Hubble's open access hatches. At right, a close-up of Smith preparing tools.

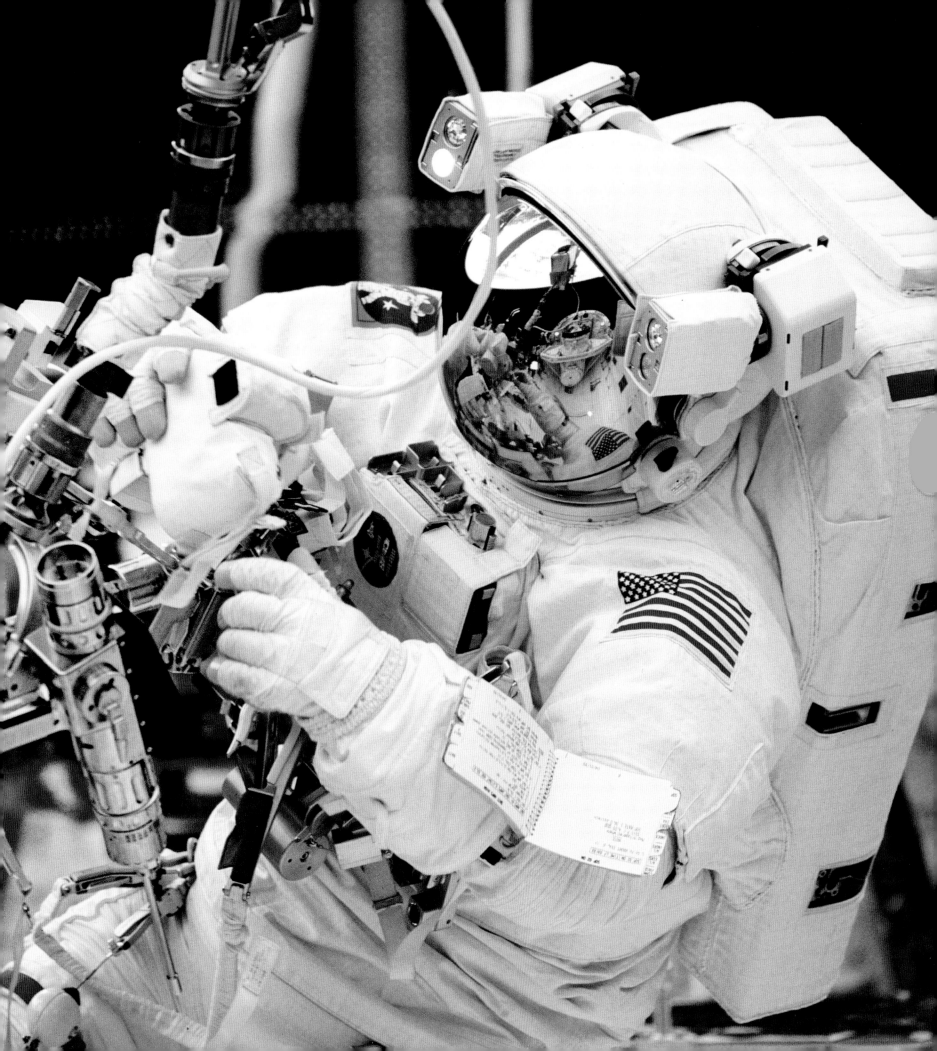

STS-99 *Endeavour*

Launched	February 11, 2000
Landed	February 22, 2000
Commander	Kevin R. Kregel
Pilot	Dominic L. Pudwill Gorie

Mission specialists
Janet L. Kavandi, Mamoru Mohri,
Gerhard P. J. Thiele, Janice E. Voss

Milestones
The shuttle radar topography mission mast was
deployed to its full length for a successful
mapping mission. *Endeavour* burned up slightly
too much propellant offsetting the slight twisting
force of the mast (induced every time the shuttle
shifted its attitude). However, enough propellant
was saved to complete the mission. Also aboard
Endeavour was a student experiment, EarthKAM,
which took three thousand digital photos of the
Earth through a flight-deck window. The NASA-
sponsored program allowed middle school
students to select photo targets and receive the
images via the Internet.

◄
*This STS-99 scan shows California's notorious San Andreas
Fault along the southwestern edge of the Mojave Desert,
north of downtown Los Angeles. The San Gabriel Mountains
fill the lower left half of the image. Various types of radar
data are combined. Stress lines in the crust are revealed by
the artificial color enhancements in this digital rendering.*

STS-101 *Atlantis*

Launched	May 19, 2000
Landed	May 29, 2000
Commander	James D. Halsell Jr.
Pilot	Scott J. Horowitz

Mission specialists

Susan J. Helms, Yuri Vladimirovich Usachev,
James S. Voss, Mary Ellen Weber,
Jeffrey N. Williams

Milestones

The third mission to the space station, during
which astronauts Voss and Williams completed
one EVA to equip the exterior of the station with
an antenna and the final parts of a Russian-built
crane, Strela. Halsell and Horowitz also fired
thruster jets while *Atlantis* was still docked,
boosting the station into a slightly higher orbit.

*Flat-panel display screens in the flight deck of Atlantis
replace many mechanical gauges and old-style cathode
ray TV tubes. The new "glass cockpit" brings the shuttle
technology up to modern standards for STS-101.*

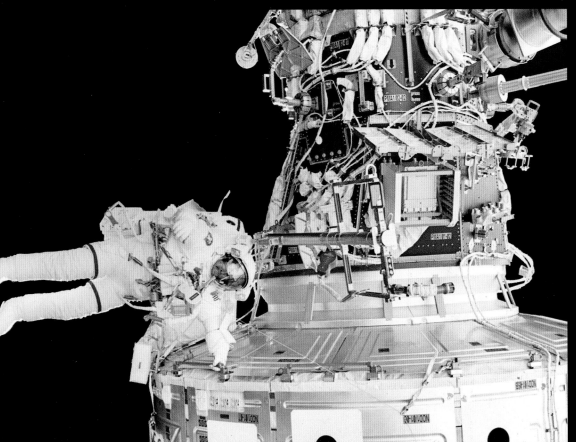

*Williams hangs onto one of the newly-installed handrails
on the Pressurized Mating Adapter during a six-hour EVA
with Voss.*

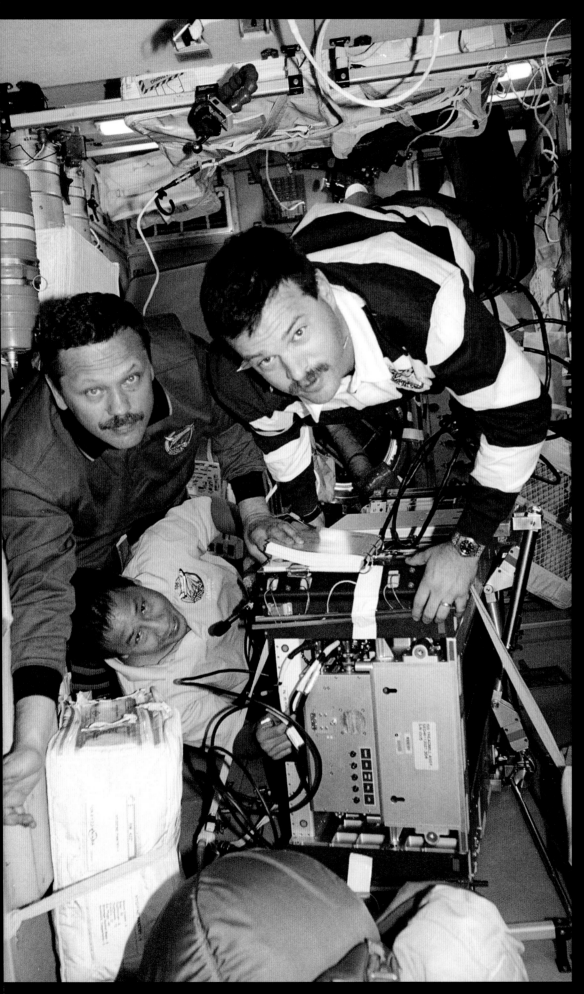

STS-106 *Atlantis*

Launched	September 8, 2000
Landed	September 20, 2000
Commander	Terrence W. Wilcutt
Pilot	Scott D. Altman

Mission specialists
Daniel C. Burbank, Edward T. Lu,
Yuri I. Malenchenko, Richard A. Mastracchio,
Boris V. Morokov

Milestones
This mission unloaded three tons of space station cargo, even as a Russian Progress supply craft already docked to the opposite end of the station was delivering its own additional load. Lu and Malenchenko conducted an EVA to connect power cables between Russian-built modules, and to install a magnetometer to serve as a compass, showing the station in respect to the Earth. Burbank and Mastracchio maneuvered their crewmates with the robotic arm.

◄
Three members of the STS-106 crew move a treadmill device. Counterclockwise from bottom are Lu, Morukov, and Altman.

Lu, captured in close-up by his spacewalking colleague, Malenchenko, although they spent most of their time working at opposite ends of the station.

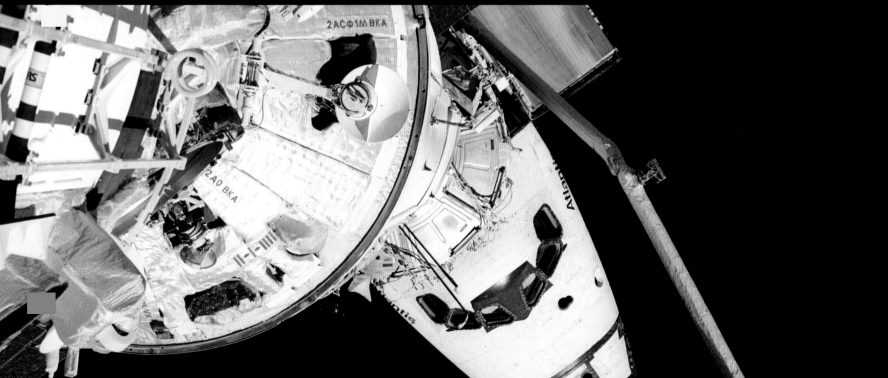

A view from the space station's perspective of STS-106 Atlantis in its docked position.

STS-92 *Discovery*

Launched	October 11, 2000
Landed	October 24, 2000
Commander	Brian Duffy
Pilot	Pamela A. Melroy

Mission specialists
Leroy Chiao, Michael E. Lopez-Alegria,
William S. McArthur Jr., Koichi Wakata,
Peter J. K. Wisoff

Milestones
During a twelve-day mission to the space station,
involving four EVAs, *Discovery*'s crew installed the
Zenith Z1 truss element (part of the large
backbone that holds the entire structure together)
and a third pressurized mating adapter for use as
a docking port for subsequent shuttle missions.

▶

*There is no such thing as "upside-down" on a space walk,
as McArthur demonstrates while holding onto a restraint
device on the robot arm. The pressurized mating adapter
is at the top of the frame.*

▶

*From the left, Melroy, Wakata, and McArthur team up for
some inflight maintenance on a temporarily extracted
part of a flight deck panel.*

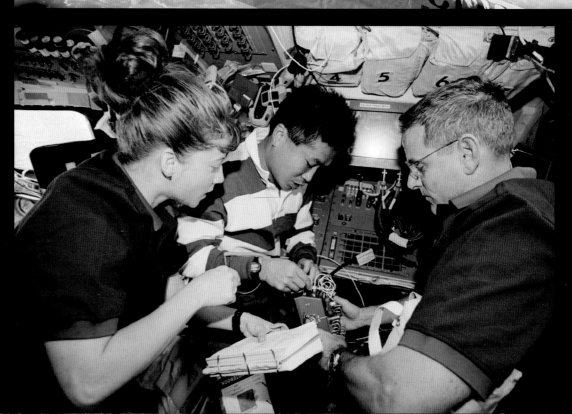

Silhouetted against the Earth, the space station elements, from the left, are the Unity node, the functional cargo block Zarya, service module Zvezda, and a Progress cargo ship.

STS-97 *Endeavour*

Launched   November 30, 2000
Landed   December 11, 2000
Commander   Brent W. Jett Jr.
Pilot   Michael J. Bloomfield

Mission specialists
Marc Garneau, Carlos I. Noriega,
Joseph R. Tanner

Milestones
Endeavour's astronauts completed three space
walks to deliver a truss element, connect an early
set of solar arrays to the space station, and
prepare a docking port for the impending arrival
of a major pressurized component, the U.S.-built
Destiny laboratory module.

*Tanner on EVA, his face hidden behind a reflective gold
vizor, with the Earth as a stunning backdrop. This type of
image has familiarized us to humans routinely working in
a new environment.*

*From the left, Shepherd (the station's first commander)
and STS-97 crew members Noriega, Bloomfield, Jett, and
Garneau get together in the Unity node.*

STS-98 *Atlantis*

Launched	February 7, 2001
Landed	February 20, 2001
Commander	Kenneth Cockrell
Pilot	Mark Polansky

Mission specialists
Robert L. Curbeam Jr., Marsha S. Ivins,
Thomas D. Jones

Milestones
The Destiny laboratory was successfully installed
on the space station using the remote manipulator
system. Ivins pulled the sixteen-ton module from
the cargo bay, then flipped it 180 degrees and
moved it into position on the station's assigned
connecting port. A set of automatic latches and
bolts tightened to hold it permanently in place.
The complex operation was backed up with EVAs
conducted by Curbeam and Jones.

▶

*The Destiny module is pulled out of STS-98's
payload bay in readiness for installation.*

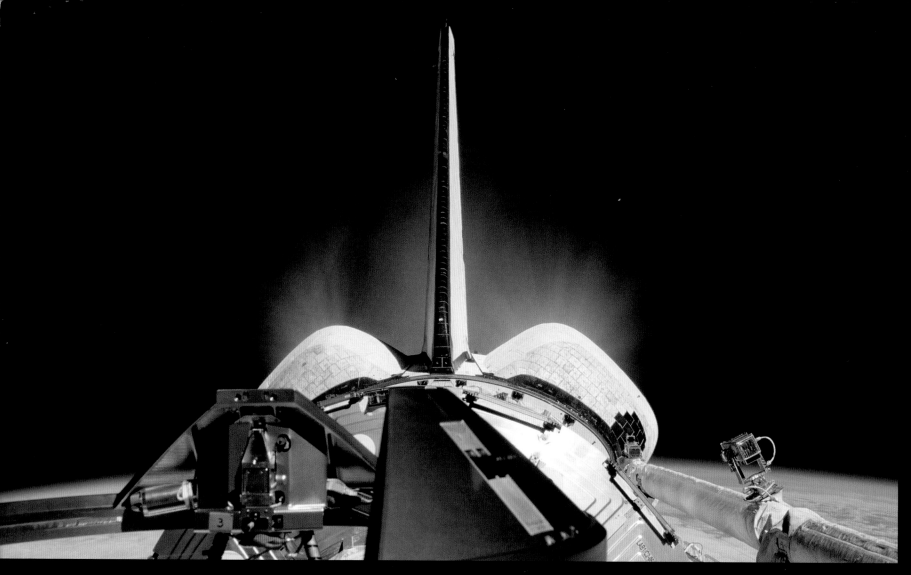

STS-98 *Atlantis conducts a thruster firing. Faint virbrations can be felt on the flight deck, but (contrary to the impression given by most space fiction movies) the firing is noiseless in the vacuum of space.*

In this abstractly beautiful shot, Atlantis moves away from the space station at the conclusion of the iSTS-98 visit.

▲
Thomas gazes through an aft window as Discovery *approaches the docking bay of the space station.*

Congratulations all round, as STS-102 Discovery *makes a clean docking with the station. From left to right are Helms, Voss, Weatherbee, and Thomas.*
▼

STS-102 *Discovery*

Launched	March 8, 2001
Landed	March 21, 2001
Commander	James D. Wetherbee
Pilot	James Kelly

Mission specialists
Paul Richards, Andrew S. W. Thomas

Station crew to orbit
Susan J. Helms, Yuri V. Usachev, James S. Voss

Station crew to Earth
Yuri P. Gidzenko, Sergei K. Krikalev,
William M. Shepherd

Milestones

A sunrise launch carried the second resident crew to the space station as well as "Leonardo," the first Multi-Purpose Logistics Module (MPLM), a European-built pressurized cylinder filled with supplies, including science racks for transfer to the Destiny laboratory. *Discovery*'s space walkers, Voss, Helms, Thomas, and Richards, installed a mounting platform on the outside of the station ahead of a future delivery, a Canadian-built robotic arm system. Helms, Usachyov, and Voss remained aboard the station as new residents, while Yuri Gidzenko, Krikalev, and Shepherd returned to Earth aboard *Discovery*.

STS-102's main payload is the Italian Space Agency-built Multipurpose Logistics Module (MPLM), called "Leonardo," and one in a family of three similar modules that serve as delivery trucks filled with equipment, experiments, and supplies for the station.

Cosmonaut Gidzenko surrounded by in-transit hardware aboard Leonardo. Gidzenko boarded the space station in November 2000, and returned to Earth with STS-102.

The Destiny module is in the foreground of this departing STS-102 view of the fast-expanding space station.

STS-100 *Endeavour*

Launched	April 19, 2001
Landed	May 1, 2001
Commander	Kent V. Rominger
Pilot	Jeffrey S. Ashby

Mission specialists

Umberto Guidoni, Chris A. Hadfield,
Yuri Lonchakov, Scott Parazynski, John Phillips

Milestones

The Canadarm2 advanced robotic arm was
delivered and attached to a pallet on the outside
of Destiny, from where it was directed to "walk"
off the pallet and grab onto an electrical fixture
that provided power and data to the arm.
Canadarm had been inside *Endeavour*'s payload
bay during launch. Subsequent exchanges of
hardware from the shuttle arm to the station's arm
represented the first ever robot-to-robot transfers
in space, although Hadfield and Parazynski made
two EVAs to assist the installation. Ordinary
human muscle power was needed for hauling
cargo out of the latest logistics module (Rafaello)
and swapping it for station trash and homeward-
bound science specimens.

*Space station and STS-100 crew members in the emptied
STS-100 cargo module. Clockwise from the nine o'clock
point in the circle are Rominger, Lonchakov, Usachev,
Guidoni, Voss, Ashby, Parazynski, Phillips, and Hadfield,
with Helms at the center.*

Parazynski during the first of two STS-100 space walks. The window on Destiny is reflected in his helmet, while he and Hadfield install a Canadian-built robot arm.

Helms and Voss, looking happily out the window of the Destiny laboratory module while Parazynski and Hadfield complete their EVA.

STS-104 *Atlantis*

Launched	July 12, 2001
Landed	July 24, 2001
Commander	Steven W. Lindsey
Pilot	Charles Hobaugh

Mission specialists
Michael L. Gernhardt, Janet L. Kavandi,
James F. Reilly

Milestones

In a series of three space walks, a new EVA airlock
module, known as Quest, was attached to the
Unity node. Station crew member Susan Helms
lifted the airlock out of *Atlantis'* payload bay using
the Canadarm2 and maneuvered it to the berthing
port on Unity, while space walkers Gernhardt and
Reilly provided additional guidance from outside.

◄
An impressive view of Atlantis, *docked to the Destiny
laboratory during the STS-104 mission.*

STS-104's primary payload, the Quest airlock, in the process of being installed onto the starboard side of the Unity node. Astronaut Helms is operating the space station's new robot arm to collect the latest delivery.

Gernhardt, in the new Quest airlock, prepares for the first EVA undertaken from the space station.

STS-105 *Discovery*

Launched	August 10, 2001
Landed	August 22, 2001
Commander	Scott J. Horowitz
Pilot	Frederick W. Sturckow

Mission specialists
Daniel T. Barry, Patrick Forrester

Station crew to orbit
Frank L. Culbertson Jr., Vladimir N. Dezhurov,
Mikhail Tyurin

Station crew to Earth
Susan J. Helms, Yuri V. Usachev, James S. Voss

Milestones
Discovery's payload included the early ammonia
servicer (part of the cooling system installed on
the outside of the station) and the Leonardo
logistics module, crammed with supplies. Barry
and Forrester completed two EVAs to install the
servicer, while Horowitz operated *Discovery*'s
robot arm and Sturckow choreographed the space
walk from the flight deck. New arrivals Culbertson,
Dezhurov, and Tyurin became the station's
resident crew, while Helms, Usachyov, and Voss
returned to Earth aboard *Discovery*.

▲
STS-105 on the pad prior to launch. The crew access arm,
tipped by the "white room," is in the foreground.

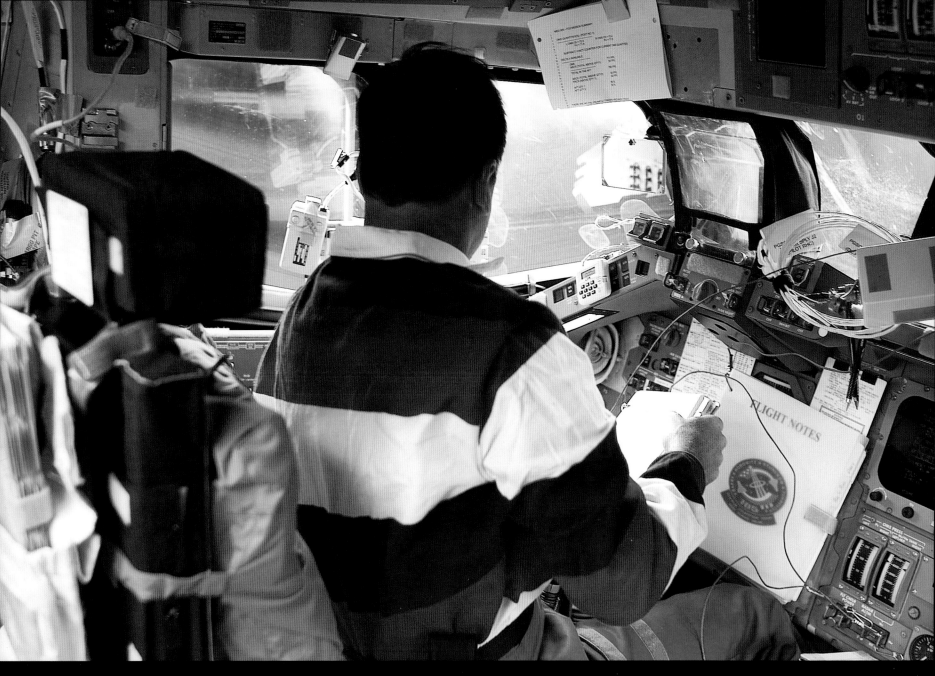

Horowitz, STS-105 commander, checks his notes on the flight deck during rendezvous operations.

The station crew's view of Discovery as it nears the final moment of docking.

Astronauts and cosmonauts share a meal in the Zvezda service module. STS-105 commander Horowitz opens a can of food as he floats near the ceiling. Others, from the left, are: Helms, Culbertson, Usachev, Voss, and Dezhurov.

Voss at a work station in the Destiny lab, while Horowitz floats through the hatchway leading to the Unity node.

Backdropped by a sunrise, a panel attached to the outside of the Quest airlock, studded with different materials, tests for "weathering" in the harsh environment of space.

Forrester waves at crewmates inside Discovery during the second EVA undertaken with Barry for STS-105.

As STS-105 heads for home, a last look at the station shows the gleaming new Quest airlock, visible at the top right of this image.

STS-108 *Endeavour*

Launched	December 5, 2001
Landed	December 17, 2001
Commander	Dominic L. Pudwill Gorie
Pilot	Mark E. Kelly

Mission specialists
Linda M. Godwin, Daniel M. Tani

Station crew to orbit
Daniel W. Bursch, Yuri Onufrienko, Carl E. Walz

Station crew to Earth
Frank L. Culbertson Jr., Mikhail Tyurin,
Vladimir N. Dezhurov

Milestones
Bursch, Onufriyenko, and Walz took residence aboard the station, while Culbertson, Dezhurov, and Tyurin took a ride home with *Endeavour*. The two spaceraft remained hard-docked during a week of cargo swapping (via the Raffaello module) and joint operations by the ten combined crew members, including an EVA by Godwin and Tani to install insulation on mechanisms to rotate the station's solar arrays. Just before returning to Earth, *Endeavour*'s crew deployed STARSHINE 2, the second such reflector target designed for use by student experimenters around the world.

Homecoming cosmonaut Dezhurov (left) and STS-108 pilot Kelly take a farewell look at the station as Endeavour *moves away after undocking.*

Dezhurov photographs his erstwhile home through Endeavour's windows.

The first quartet of the station's main solar power arrays are turned hungrily towards the sun. At the left of the complex, Canadarm2, the station's own robot arm, is visible

▲ *Linnehan and Grunsfeld (partially obscured behind Linnehan) replace Hubble's starboard solar array.*

STS-109 *Columbia*

Launched	March 1, 2002
Landed	March 12, 2002
Commander	Scott D. Altman
Pilot	Duane G. Carey

Payload commander
John M. Grunsfeld

Mission specialists
Nancy J. Currie, Richard M. Linnehan,
Michael J. Massimino, James H. Newman

Milestones
The eleven-day mission rejuvenated the Hubble Space Telescope in a series of five space walks. After grasping the telescope and pulling it into the payload bay, space walkers Grunsfeld, Linnehan, Massimino, and Newman, assisted by Currie operating the shuttle's robotic arm, installed new power equipment and optics.

▶ *Backdropped by desert, Hubble is the subject for an almost painterly STS-109 portrait of the world's most popular science instrument.*

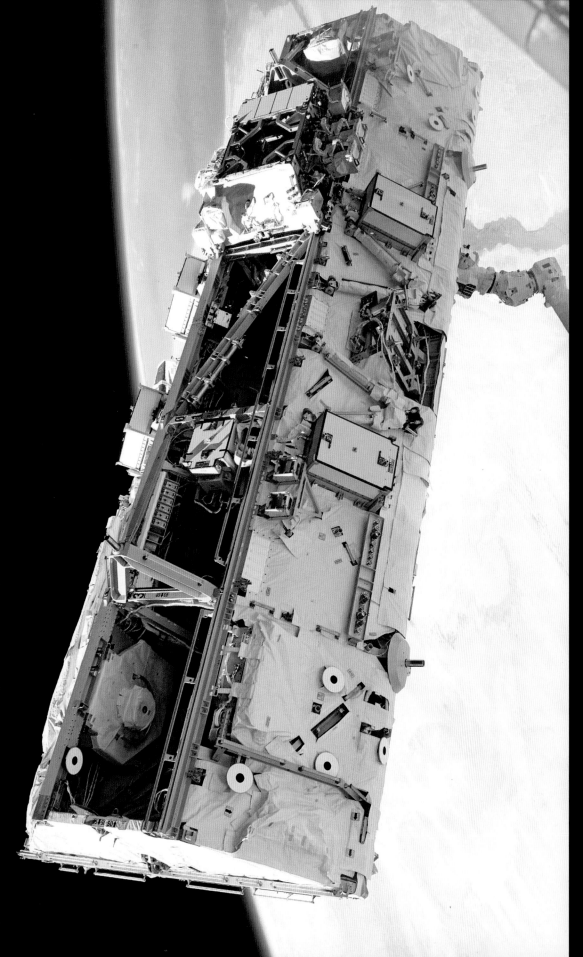

The huge S0 (S-Zero) truss element is carefully removed from the cargo bay during STS-110.

►

Morin totes an S0 (S-zero) keel pin: one of countless pieces of hardware whose names and purposes make sense only to space station experts.

STS-110 *Atlantis*

Launched	April 8, 2002
Landed	April 19, 2002
Commander	Michael J. Bloomfield
Pilot	Stephen N. Frick

Mission specialists

Lee M. E. Morin, Ellen Ochoa, Jerry L. Ross, Steven L. Smith, Rex J. Walheim

Milestones

Installation of the Space Station's S0 (S-zero) truss element was the primary objective. Ochoa lifted it out of the cargo bay with the station's robotic arm and maneuvered it onto a clamp at the top of the Destiny module. The truss contained navigational devices, computers, and cooling and power systems needed to attach additional trusses, solar arrays, and modules. Four space walks were required to complete all the tasks.

▲ *STS-110 space walker Walheim is backdropped by a pair of the station's vast solar panel arrays.*

◄ *Ochoa looks through the Earth observation window in the Destiny laboratory, where she gets a clear view of STS-110's tailplane.*

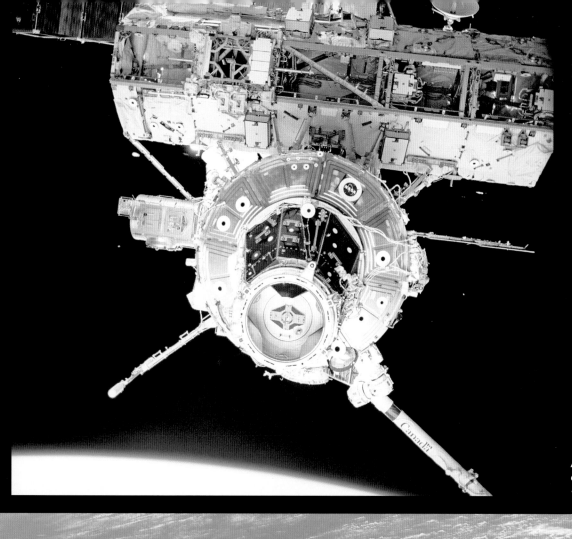

STS-110 Atlantis undocks from the station and embarks on a slow fly-around inspection, taking in this view of the docking port and recently installed truss.

A wider view of the station as Atlantis prepares to leave orbit and head for home.

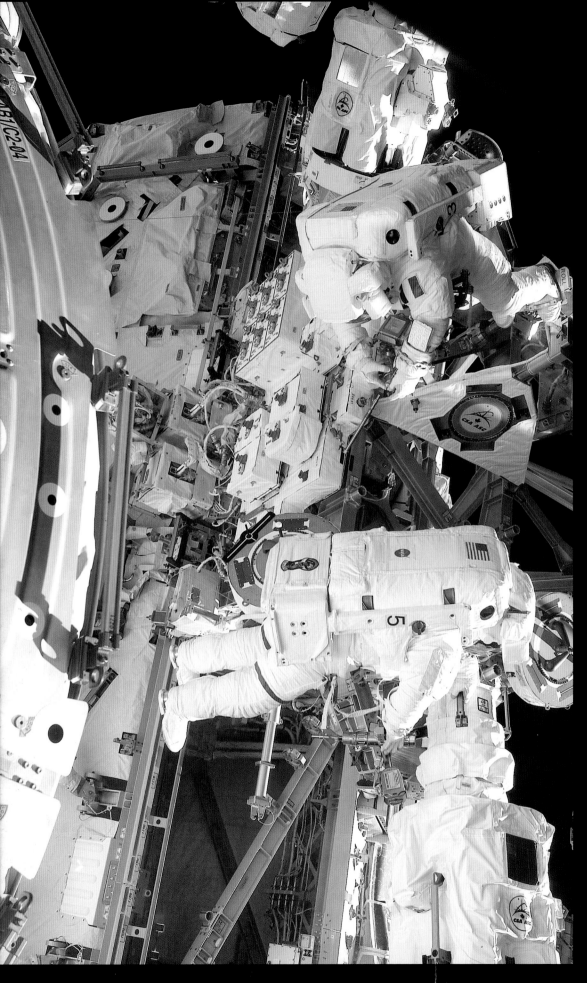

STS-111 *Endeavour*

Launched	June 5, 2002
Landed	June 19, 2002
Commander	Kenneth D. Cockrell
Pilot	Paul Lockhart

Mission specialists
Franklin R. Chang-Diaz, Philippe Perrin

Station crew to orbit
Valery Korzun, Sergei Treschev, Peggy Whitson

Station crew to Earth
Daniel Bursch, Yuri Onufriyenko, Carl Walz

Milestones
Cockrell moved the Leonardo module from
Endeavour's payload bay to the Unity module.
The crew then transferred cargo to the station,
including a new science rack for microgravity
experiments and a glovebox that allowed station
crews to conduct biological experiments requiring
careful isolation. The station's Expedition 4 crew,
Bursch, Onufriyenko, and Walz, ended their 182-
day residence, and the Expedition 5 team,
Korzun, Treschev, and Whitson, took over.

◄
*Chang-Diaz and Perrin (top) attach power, data and video
cables to a mobile transporter platform , a device like an
electric railroad cart on tracks that assists in moving
equipment along the outside of the station.*

STS-111 snapped the Manicouagan Reservoir in Quebec, Canada: the site of an impact crater, 60 miles (100 km) wide, which formed 212 million years ago when a meteorite crashed into this area.

STS-112 pilot Melroy on the flight deck of shuttle Atlantis on day two of the mission.

STS-112 *Atlantis*

Launched	October 7, 2002
Landed	October 18, 2002
Commander	Jeffrey S. Ashby
Pilot	Pamela A. Melroy

Mission specialists
Sandra H. Magnus, Piers Sellers, David Wolf, Fyodor Yurchikhin

Milestones
The largest payloads were the S1 integrated truss segment and the Crew and Equipment Translation Aid (CETA), the first of two human-powered carts that ride along a rail on the station's exterior, providing mobile work platforms for space walking astronauts. In the course of three EVAs, astronauts installed the fourteen-ton S1 truss to the station. Less dramatic interior duties included repairs to the station's exercise treadmill.

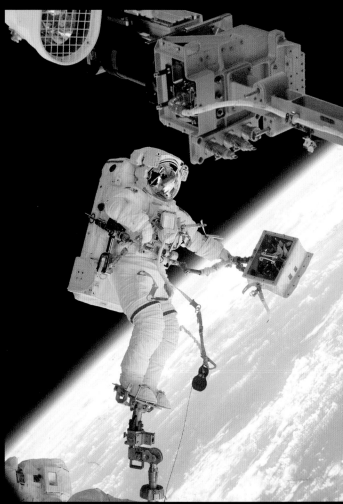

Atlantis approaches the station, its payload bay loaded with truss hardware and other equipment.

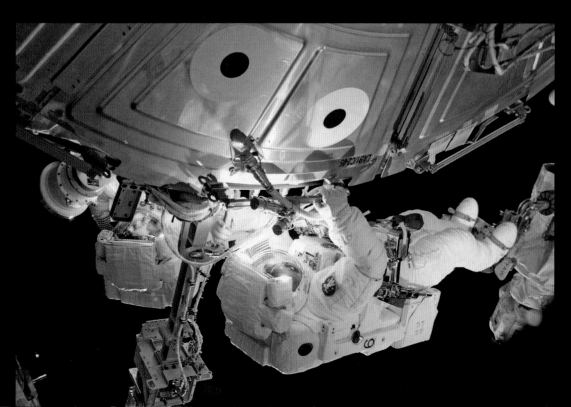

Wolf on the tip of the Canadarm2 attached to the station, during one of the EVA's associated with delivery of the second truss element.

STS-112 astronaut Sellers uses the lamps on his helmet to light the scene while working on the night side of Earth.

STS-113 *Endeavour*

Launched   November 23, 2002
Landed   December 7, 2002
Commander   James D. Wetherbee
Pilot   Paul Lockhart

Mission specialists
Michael E. Lopez-Alegria, John Herrington

Milestones
Over the course of the two-week mission, the STS-113 crew and the station's crew worked together on three EVAs to install an additional truss component with the uncanny help of robot arms on both of the docked spacecraft.

▶
Herrington waves, with the nose of STS-13's Endeavour shuttle behind him.

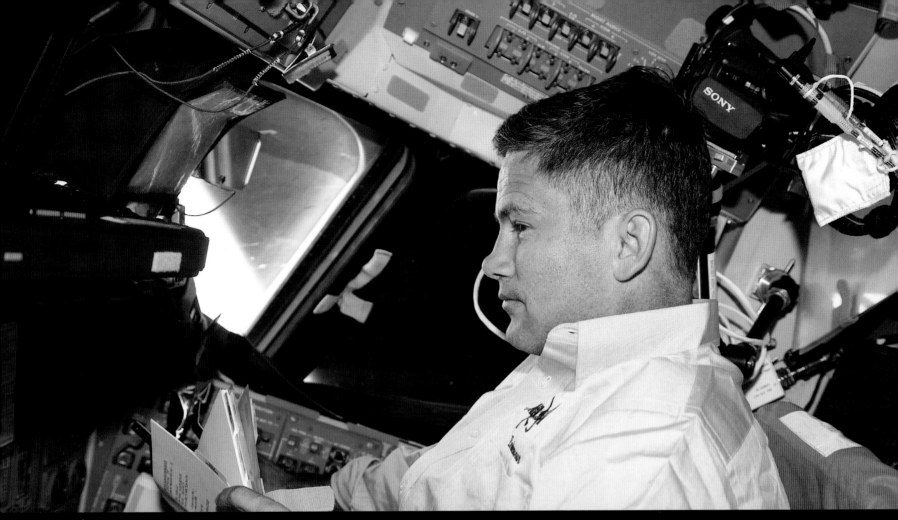

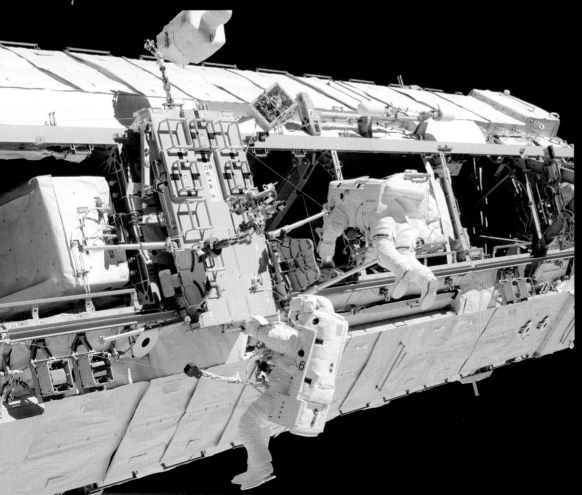

Seated at his pilot's station, Lockhart looks through a checklist as part of his workload on STS-113.

Herrington (left) and Lopez-Alegria work on one of the station's massive, and recently installed, truss structures.

AT THE
CROSSROADS

RESHAPING THE SPACE PROGRAM

AT THE CROSSROADS
RESHAPING THE SPACE PROGRAM

The era of the space shuttle has ended. We look back on dramas, tragedies, high costs, and potentially better types of spaceship that should have been built, but never were. We also look back on an era of genuinely great achievements: some of our boldest and most ambitious aspirations for space were, indeed, fully realized. The question is this: what next?

TAKE A LOOK AT a typical copy of a science fiction magazine from half a century ago, or revisit Wernher von Braun's *Collier's* articles, and it would be hard to distinguish the reality of recent events in space from many of the predictions held out for us by the first great rocket visionaries. The media likes to tell us that the Space Age ended with the final Apollo missions to the Moon, yet the stunning pictures regularly returned by shuttle crews and space station occupants at the dawn of the twenty-first century told a different story. We really did seem to be on the verge of becoming a space-faring civilization. But then, on the first day of February 2003, NASA repeated its earlier mistake of allowing acknowledged flaws in the shuttle's existing hardware to be regarded as acceptable for flight. Yet again, the agency forgot the essentially experimental nature of the system and fell into its old habit of treating it as an operational vehicle—and yet again, on the shuttle system's 113th flight, Nature refused to be fooled.

Returning from a microgravity science expedition designated STS-107, *Columbia* disintegrated during re-entry at the end of an otherwise flawless sixteen-day mission. The entire crew of seven were lost: Commander Rick D. Husband, pilot William C. McCool, payload commander Michael P. Anderson, payload specialist Ilan Ramon, and Mission specialists Kalpana Chawla, David M. Brown, and Laurel Clark. The *Columbia* Accident Investigation Board (CAIB), chaired by a respected aerospace executive and former Lockheed Martin president Norman Augustine, made an unforgiving analysis over the following months and concluded that a suitcase-sized chunk of foam insulation from the external fuel tank had peeled off during launch, as had

been indicated by film of the launch viewed the day after liftoff. It slammed into the front of *Columbia*'s left wing, making a small but ultimately lethal hole in the heat-resistant panels on the leading edge. The re-entry of a shuttle is usually one of its safest and best-understood procedures, but, as with the *Challenger* explosion seventeen years earlier, errors in managerial judgment had led to an incorrect assessment of the vehicle's condition. A number of requests from engineers for an astronaut inspection or satellite imaging to assess possible damage to the heat shields (as had been done during *Columbia*'s first mission over twenty years before) were denied by management.

Thermal insulation foam protects liquid oxygen and hydrogen inside the external fuel tank from warming up. The foam was sprayed on, rather like wall cavity filler, and then set hard. It was very lightweight and thermally efficient, but engineers discovered that tiny pockets of air and moisture tended to become trapped between the foam and the tank's metal shell. During the stresses of launch, the pockets warmed up and expanded, forcing imperfectly attached pieces of foam to peel away. NASA managers thought that the risks of foam impact damage were slight, based on the fact that no shuttle flight up until then had suffered significant damage. The devil was in that notion of "significant." Minor dings and dents on the orbiter's tile insulation, made by impacts from small flecks of falling tank insulation, were common to almost all shuttle missions, so much so that they were assumed to be unimportant, because none had ever compromised flight safety. It is hard to conceive of any vehicle in other realms of transport where the regular disintegration of

Launch pad 39B at the Kennedy Space Center: one of two flame-scoured veterans of the Apollo lunar missions, and of the space shuttle era. What will be their future?

components and impacts of broken pieces on other parts of the system would be tolerated. It is equally improbable to imagine that with the stakes so high and a previous disaster as an object lesson, managers eschewed the gathering of actual evidence in favor of reviewing statistical probabilities—as if to say, "Maybe you saw smoke, but why check if your house is on fire when it's never been on fire before?"

The ultimate truth is that rocket flight is, and always has been, inherently risky. Up to his retirement in 2004, John Young, still pounding the corridors of NASA's Johnson Space Center in his mid-seventies, liked to greet all new astronaut candidates with a flip chart

presentation about the hazards of their chosen profession. Even those coming from a military aviation background were somewhat startled by his frankness. "You have a one-in-258 chance of a catastrophic failure on any given shuttle mission," he would tell them. "Flying one shuttle mission is as dangerous as any combat missions you would fly." In fact, by the time Young made that assessment, the failure rate was one in 53, based on the cold, hard fact that, by February 2003, two missions out of 113 had come to grief.

As the shuttle passes into history, tough decisions lie ahead. Will U.S. leadership in space be continued, or must this vision be consigned to museums, like this once-mighty Saturn V?

empty tanks or boosters and return to Earth as an unpowered glider just like the current shuttle. A larger version of X-33, known as VentureStar, was set to revolutionize space travel, delivering astronauts or cargo alike in sealed pods with no need for a crewed cockpit at the front, an uncomfortable idea for astronauts used to being in the driver's seat. Unfortunately for Lockheed, X-33's experimental lightweight carbon composite fuel tanks were too ambitious and could not hold their liquid hydrogen fuel without developing leaks. Lockheed planned to fit simpler aluminum tanks instead, but the additional weight made X-33 too heavy to reach space. The promising but overambitious project was cancelled in 2001. What had seeemed a promising line of research became just one of many paths not taken.

Back to basics

On a pragmatic technical level, the CAIB concluded that it is too dangerous to carry astronauts in the same element of a spacecraft that contains the rocket engines because of the risk of explosions, debris collisions, or aerodynamic break-up of the structure. The board noted that the 1960s Apollo capsules were remarkably safe overall. The tough, compact crew compartment could always be instantly separated from other modules or booster rockets. When a section of the Apollo 13 spacecraft exploded on the way to the moon in April 1970, the rear service module with the rocket engine was torn open, but the crew capsule itself was unharmed and eventually returned safely to Earth. The shuttle had no such self-contained crew module capable of cleanly separating from the rest of the machinery in the event of a problem.

The CAIB also recommended that future crew modules for any spacecraft should be carried on the uppermost tips of their carrier rockets (what's known as an in-line arrangement) so that no launch debris can fall onto sensitive modules from any hardware above, as had happened in the Columbia disaster. Finally, the CAIB insisted that NASA's remaining shuttles, Atlantis, Endeavor, and Discovery, should not push their luck beyond what was necessary to complete the assembly of the International Space Station. They should be retired from service by the year 2010.

The U.S. president at that time, George W. Bush, concurred. His administration worked with NASA to conceive the Constellation program, a return to the Moon using launch hardware adapted from the shuttle's SRBs, augmented by a new upper stage: a conical crew capsule known as Orion and a separate liquid-fueled heavy-lift launcher for other Moon-bound cargo, such as a landing craft.

In the summer of 2009, as NASA celebrated the fortieth anniversary of Apollo 11's lunar landing mission, a new president, Barack Obama, was faced by a global financial crisis unprecedented in modern times. While he augmented NASA's annual budget, he and his advisors did not feel that the American public would be interested in seeing astronauts go to the Moon again, merely repeating what had been so superbly accomplished already by Apollo 11 in 1969. The White House favored killing off the Orion project, while redirecting NASA to invent new technologies for possible crewed missions to asteroids rather than the Moon. Congress was not entirely in agreement. A year later, after many protracted arguments, in which even famous-name astronauts were pitted against each other in the press, Congress voted to continue development of the Orion crew capsule and to maintain much of the intellectual and personnel legacy from the shuttle program rather than dispensing with it completely. Even so, the shuttle orbiter fleet was moved closer toward retirement in favor of new heavy-lift rockets, and NASA was directed to develop radical propulsion systems and other technologies to enable the next generation of human adventures in deep space, driven both by NASA and by private space launch entrepreneurs. We now await the judgment of history on these decisions.

Nothing uniquely dangerous

Only someone totally dismissive of the professionalism, goodwill, and engineering excellence within NASA would suggest that the agency was recklessly careless, either in the case of the Columbia's loss or the Challenger explosion. As the shuttle era draws to a close, it's time, perhaps, to put these two technologically defining accidents into a more forgiving perspective.

Consider the amazing reliability of another defining piece of American aerospace technology: Boeing's familiar 747 jumbo jet airliner. For some years before the first models entered commercial service, prototypes lumbered up into the air for hundreds of test flights. Even those first takeoffs were preceded by repetitive sessions where the wheels never left the ground. There were runway speedups and landing gear braking tests, just to make sure that the plane could behave itself under the dullest conditions imaginable: trundling along the runway or rolling to a halt by the embarkation gate. Then, in the air, pilots time and again pushed the plane to the edges of its stalling speeds, making sure that the aerodynamics conformed to expectations. As a consequence of this test flight regime, by the time that the 747 entered service, its designers had a pretty thorough understanding of how the machine should be

Shuttle: the next generation

Throughout the 1990s NASA had tried to develop a fully reusable and cost-effective replacement for the shuttle. McDonnell Douglas came up with Delta Clipper, a vehicle that lifted off and landed vertically like a classic science fiction spaceship. It relied on powerful braking rockets for touchdown. A quarter-scale test version of Delta Clipper confirmed its critics' worst fears by catching fire as soon as it reached the ground.

In July 1996 NASA opted instead for Lockheed Martin's X-33 lifting body design, with a stubby triangular fuselage and almost no need for conventional wings. It would launch as a single unit without shedding flown, operated, and serviced. By contrast, all the shuttle missions that have ever been flown have to be considered essentially as a series of test flights, because no one in the history of rocketry has ever flown *anything* into space and back so many times that its every last quirk and snag has been identified and ironed out. The shuttle's flight log of 135 launches constitutes a reasonable tally for a test flight regime, not an operational record from which reliable safety conclusions and performance data can be drawn. The O-rings had *seemed* to be capable of operating within safe limits because until the *Challenger* explosion none had actually failed. The frequent dents and dings found in thermal insulation at the end of shuttle missions were so common, they *seemed* not to matter . . . except that worrying anomalies and partial failures had been reported, time and again, by people whose voices weren't getting through to NASA's upper echelons.

If the physicist Richard Feynman was the rebel outsider during the 1986 *Challenger* investigation, then *Columbia*'s out-of-school interrogator was Diane Vaughan, the sociologist renowned for her exhaustive 1996 book *The Challenger Launch Decision*. For her, "the causes of the two accidents were identical. In both cases, they got to the point that flying with known flaws was normal and acceptable. Of course, after the accident, they were shocked and horrified as they saw what they had done." Vaughan coined the phrase "normalization of deviance" to cover these disturbing misjudgments. Yet she was always at pains to point out that individual shuttle managers were not to blame. "After getting deeper into the data, it turned out they had not violated rules at all, but had actually conformed to all NASA requirements."

The stark truth was that the shuttle had made too few flights for any of the reliability numbers or risk assessments to be accurate—or, to put it another way, the figures were sufficiently unreliable that they could be reinterpreted more or less endlessly. Given the huge social, political, and budgetary pressures on NASA, the "Go for launch" votes usually won. "The reality was that every attempt they made to quantify and clarify risk gave them no help," Vaughan says, "because there were so many thousands of components on the shuttle."

Would our individual judgments have been any better, given the thousandfold pieces of data that had to be checked every time a shuttle flew? Perhaps, one day, we will be able to boast space vehicles as reliable and well understood as any ordinary airplane. That time is not yet come, because we are still in the earliest days of our space-faring civilization and still dependent on complex, costly, and barely contained infernos of furious

chemical energies to get us off the planet. Dr. John Logsdon, former head of the Space Policy Institute in Washington D.C. and a veteran of several inquiries, post-mortems, and evaluations of the American space program, including the CAIB, reminds us that there has never been anything uniquely dangerous about the shuttle. "Naturally we had an emphasis on investigating the specific problems that doomed *Columbia*, but space vehicles are risky across the board, and they need constant vigilance to be flown successfully." Most astronauts agree with that sentiment—and they are the people, after all, who take the sharp end of the risks. Thomas D. Jones, a veteran of four shuttle flights, speaks for all his colleagues in saying, "An astronaut makes a conscious choice to repeatedly venture into a harsh environment to gain new knowledge. He or she must confront the fact that the risks of spaceflight are real and ever-present."

Ever-present indeed, but perhaps not so unacceptable as NASA's critics have so often claimed. Examining the shuttle's record beyond the noise and drama of simplistic headlines, we see a crewed spacecraft fueled with the energy equivalent of a small nuclear bomb that completed 133 flights out of 135 successfully. Contrary to what the public and media might think, the shuttle had the most reliable record of *any* space launch vehicle yet flown, amounting to a more than 97 percent success rate.

If anything, the shuttle's drawbacks stemmed from economic factors rather than safety considerations. Taking into account the actual $35 billion development costs, and adding the operational costs over the shuttle's thirty-year history, each flight has actually worked out at around $1.4 billion in real terms, or $450 million per flight if development costs throughout the first decade are set aside. It took at least eight thousand people on permanent staff to service the shuttle system. The dream of an all-purpose winged rocketship, flying cheaply and reliably like a commercial cargo plane, never quite came to fruition. Whenever a shuttle carried a satellite or a space station module in its cargo bay, the costs of launching those payloads were tremendously exaggerated because astronauts always came along for the ride. Quite apart from the massive additional complications of life support, the launch weight given up to the crew cabin—and the shuttle's heavy wings, plus all the landing gear and other equipment required for its safe return to Earth—had to be subtracted from the cargo capacity. It is unlikely, therefore, that we will ever again see humans and raw cargo carried in the same space vehicle, because it makes little economic sense. In future, splitting the hardware from the

astronauts will mean that humans can be sent aloft in small, lightweight rockets while cargo goes up on larger but simpler expendable vehicles. Humans and their payloads can then be reunited by orbital docking, just as Wernher von Braun suggested back in the 1950s.

A legacy of honor

So what went right? What were the characteristics of the shuttle that made this NASA's flagship vehicle for three decades? What enabled it to carry out so many different missions, from microgravity medical experiments to deep space probe deployments or the building of the gigantic space station? In essence, its virtues were precisely those complications that made it such a difficult and costly machine to design and operate in the first place. It had a useful "cross range," an ability to alter course during re-entry and land at a site well away from its initial orbital path. Switching between Florida or California runways gave payload managers increased flexibility when choosing orbits, while at the same time providing a margin of safety against unpredictable weather on homecoming day. The shuttle could also range in orbital altitude, reaching anywhere from one hundred to four hundred miles, thus suiting it for docking with low-orbiting space stations or high-orbiting space telescopes. It could carry anywhere between two and seven people into space and bring them home to a pinpoint runway landing, an essential tool for rotating space station crews. The winged orbiter was indeed reusable; the cargo bay was cavernous and the five linked computers on the flight deck were superb examples of rugged, reliable software and hardware design.

Throughout its collective career, the shuttle fleet placed more than three million pounds (1.36 million kilograms) of cargo into orbit and more than six hundred astronauts. By any standards, this machine clocked up an impressive record. The shuttles were exactly what NASA always said they would be: the world's first true spaceships. As we bid farewell to these fabulous flying machines, and mourn those who sacrificed their lives during the program, we should be glad to confirm their status as magnificent examples of American ingenuity and achievement.

STS-107 *Columbia*

Launched	January 16, 2003
Commander	Rick D. Husband
Pilot	Willie C. McCool

Payload commander
Michael P. Anderson

Mission specialists
David M. Brown, Kalpana Chawla, Laurel Clark

Payload specialist
Ilan Ramon

Loss of the vehicle and crew

Columbia lifted off on time on the first shuttle mission of the year. It carried seven crew members, including the first Israeli astronaut (Ramon) heading for a marathon international scientific research flight, a migrogravity research mission using SPACEHAB components as working space. On the first day of February, after a perfectly smooth sixteen-day orbital mission, *Columbia* and its crew were lost during re-entry over East Texas at about 9 a.m, just sixteen minutes before they were expected to touch down on the runway at the Kennedy Space Center. A seven-month investigation followed, including a four-month search across Texas to recover debris. A suitcase-sized chunk of foam insulation from the external tank had peeled off during launch. It slammed into the front of the left wing, making a small but lethal hole in the heat-resistant panels.

The presidentially mandated *Columbia* Accident Investigation Board (CAIB) concluded that, quite apart from all the technical and procedural fixes that would be needed to prevent such impacts in the future, the shuttle was inherently flawed because it lacked adequate crew escape systems. The CAIB concluded that the system should be removed from flight status by the end of the year 2010 and replaced with other launch vehicles. In the end, missions were permitted until June 2011, after which the orbiter fleet was retired forever.

The crew of STS-107. Seated in front are Rick D. Husband, mission commander; Kalpana Chawla, mission specialist, and William C. McCool, pilot. Standing are (from the left) David M. Brown, Laurel B. Clark, Michael P. Anderson, and Ilan Ramon (representing the Israeli Space Agency).

The wheeled rover Spirit, now on the surface of Mars, bears this memorial panel in honor of STS-107's crew

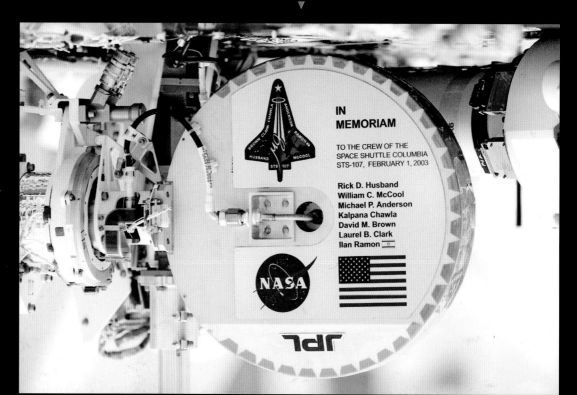

Chawla, seen here on the flight deck during the orbital phase of the mission, was the first Indian-born woman to fly into space.

Clark looks through an overhead window on the aft flight deck of Columbia. This picture was among those on a roll of unprocessed film later recovered from the debris.

STS-126 *Endeavour*

Launched	November 14, 2008
Landed	November 30, 2008
Commander	Christopher J. Ferguson
Pilot	Eric A. Boe

Mission specialists
Stephen G. Bowen, Robert S. Kimbrough,
Donald R. Pettit, Heidemarie M. Stefanyshyn-Piper

Station crew to orbit
Sandra H. Magnus

Station crew to Earth
Gregory E. Chamitoff

Milestones

This flight had long been planned as the one that would give the space station the ability to support twice the three-strong crew currently maintained on board. It also became a mission to ensure that sufficient power would be available to support those extra crew members. The faulty rotary joint on the starboard solar panel array was now intolerable, because the panels were misdirected out of the sunlight for much of the time and the power shortage was beginning to be a problem. Astronauts cleaned and lubricated the joints and replaced most of the antifriction bearings inside the mechanism. One was replaced during the STS-124 mission in June 2008.

By this point in the station's history, the public had gotten used to flawless EVAs. It was easy to forget how difficult these performances could be for time-pressed astronauts working at the limits of physical endurance. About halfway into one of the EVAs, a lubricant grease gun discharged into a kit bag. One thing led to another, and the entire bag of tools drifted away. Fortunately, a spare set of tools was on hand.

Launched	March 15, 2009
Landed	March 28, 2009
Commander	Lee Archambault
Pilot	Tony Antonelli

Mission specialists

Richard Arnold, Joseph Acaba, John Phillips, Steven Swanson,

Station crew to orbit

Koichi Wakata

Station crew to Earth

Sandra H. Magnus

Milestones

The final set of solar panels delivered by this mission included huge arrays amounting to a total "wingspan" of 240 feet, taking into account the equipment that connected the two halves and allowed them to twist as they tracked the sun. The station's solar panels were designed to generate enough electricity to power the equivalent of more than forty average homes.

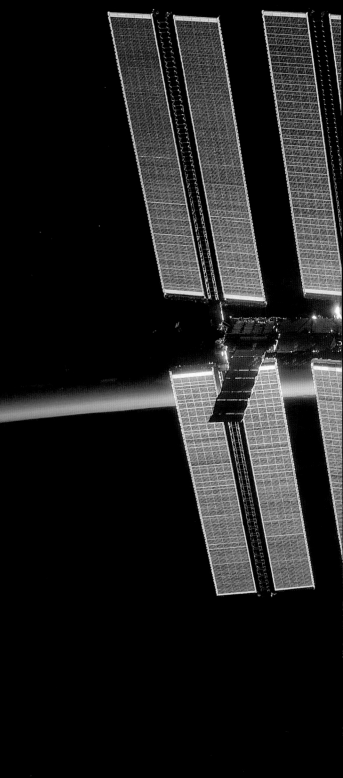

Antonelli and his crewmates captured superb views of the station as STS-119 drifted away before heading home.

▼ ▶

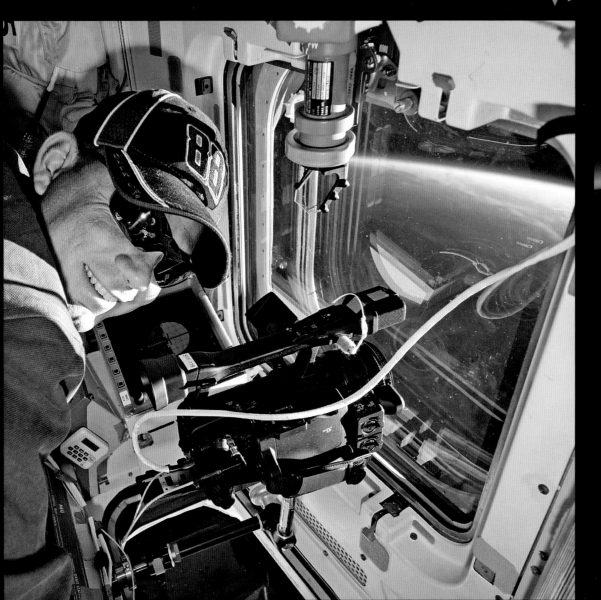

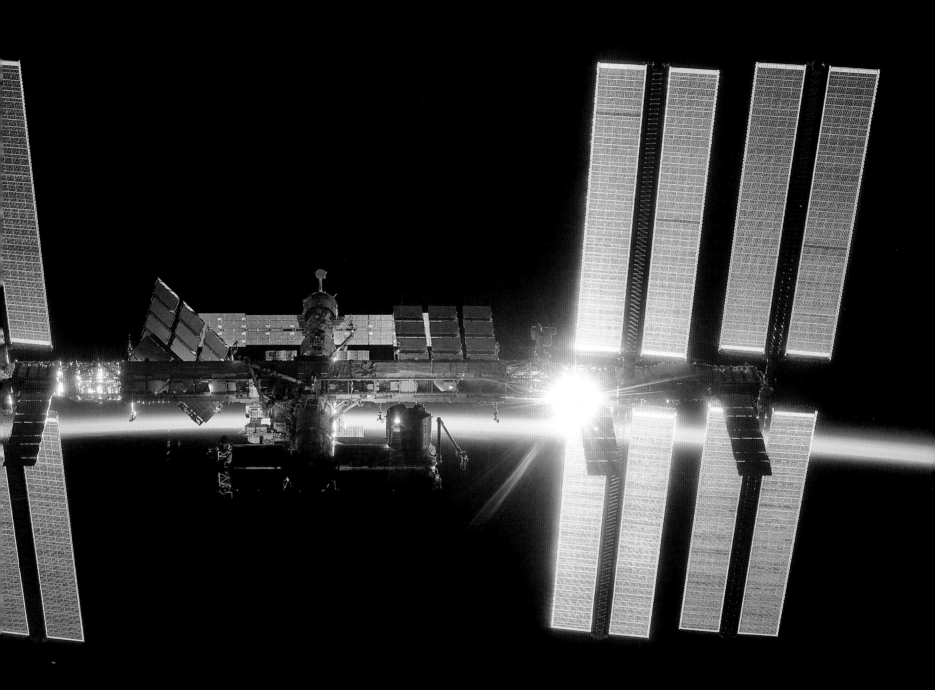

▲ McArthur (foreground) and Massimino work the robot arm controls on the aft flight deck as Hubble is captured.

◄ Hubble is overlayed with internal reflections of flight deck lights and controls in this view from the aft windows.

STS-125 *Atlantis*

Launched	May 11, 2009
Landed	May 24, 2009
Commander	Scott D. Altman
Pilot	Gregory C. Johnson

Mission specialists

Andrew J. Feustel, Michael T. Good,
John M. Grunsfeld, Michael J. Massimino,
K. Megan McArthur

Milestones

The final visit to the Hubble Space Telescope. During five EVAs, astronauts installed two new instruments, replaced failing gyroscopes and batteries, added new thermal insulation panels, and extended the instrument's operating life yet further. From this point onward, Hubble was on its own: riding on two decades of scientific success and ready for more as *Atlantis* released it back into solitary orbit for the last time.

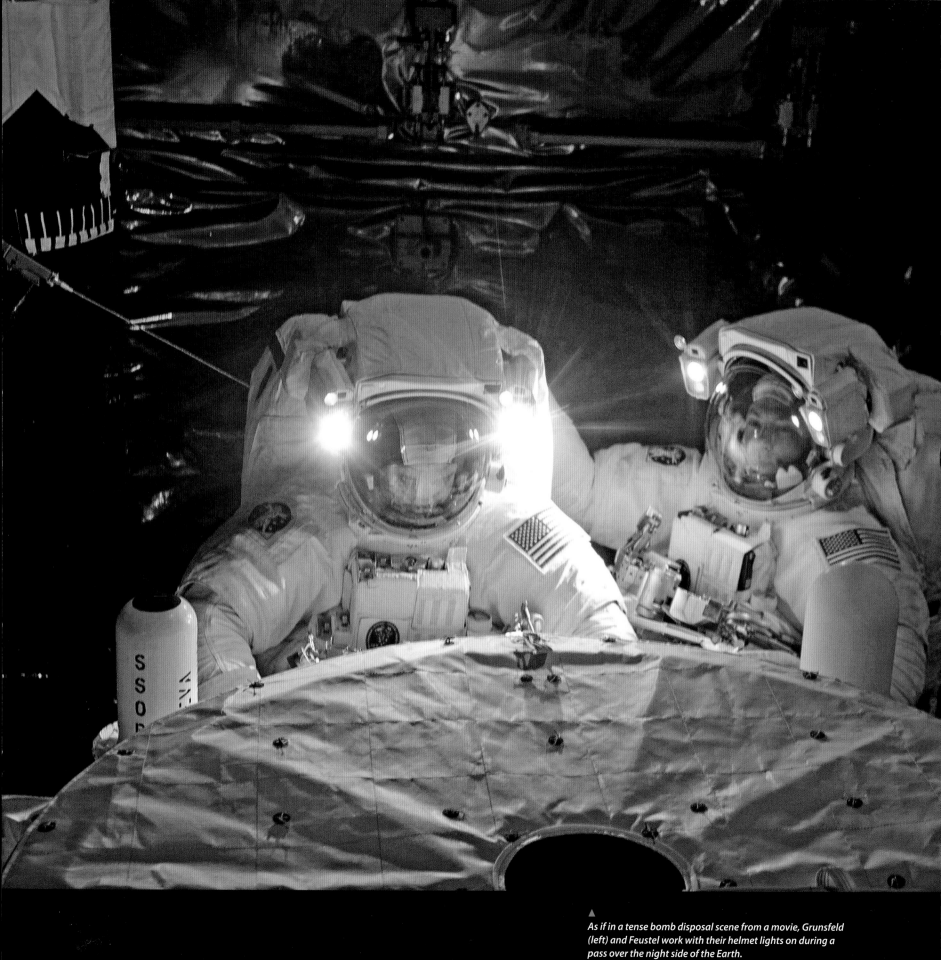

As if in a tense bomb disposal scene from a movie, Grunsfeld (left) and Feustel work with their helmet lights on during a pass over the night side of the Earth.

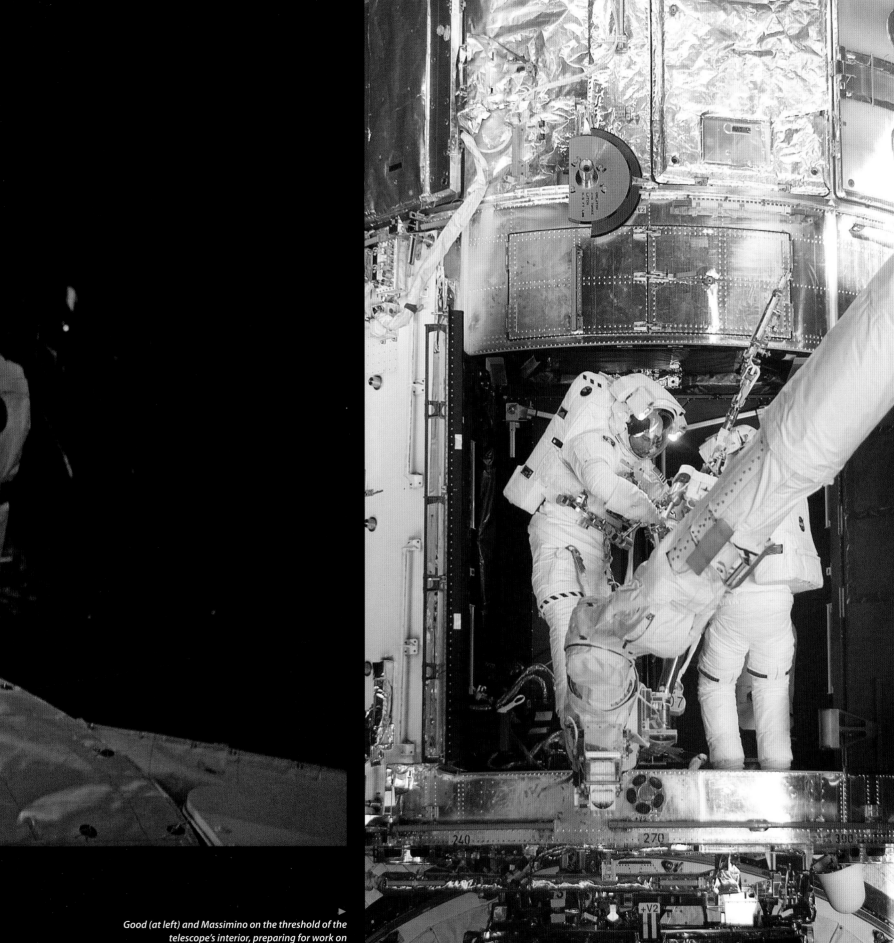

Good (at left) and Massimino on the threshold of the telescope's interior, preparing for work on the Imaging Spectrograph.

STS-127 *Endeavour*

Launched	July 15, 2009
Landed	July 31, 2009
Commander	Mark Polansky
Pilot	Doug Hurley

Mission specialists
Christopher Cassidy, Julie Payette,
Tom Marshburn, David A. Wolf,

Station crew to orbit
Timothy L. Kopra

Station crew to Earth
Koichi Wakata

Milestones
The main event was the installation of the
Japanese Experiment Module Exposed Facility, an
outdoor "porch" with its own robotic servicing
arm, for attachment to the exterior of Japan's
pressurized Kibo laboratory module. The mission
included five space walks.

◄
*The shuttle orbiter's white fuselage and wings can be hard
to photograph well in the bright Floridian sunshine, but this
shot of STS-127 on the pad is flawless.*

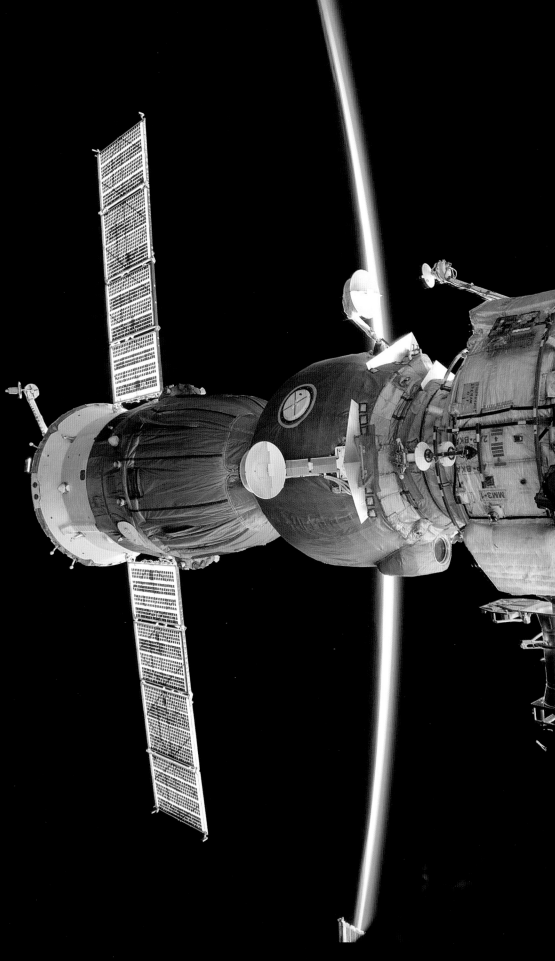

▶
*An STS-127 photo of the only alternative vehicle capable,
at that time, of sending humans into space: a Soyuz craft,
docked to the station, seen here just as the sun begins to
climb above the Earth's atmosphere.*

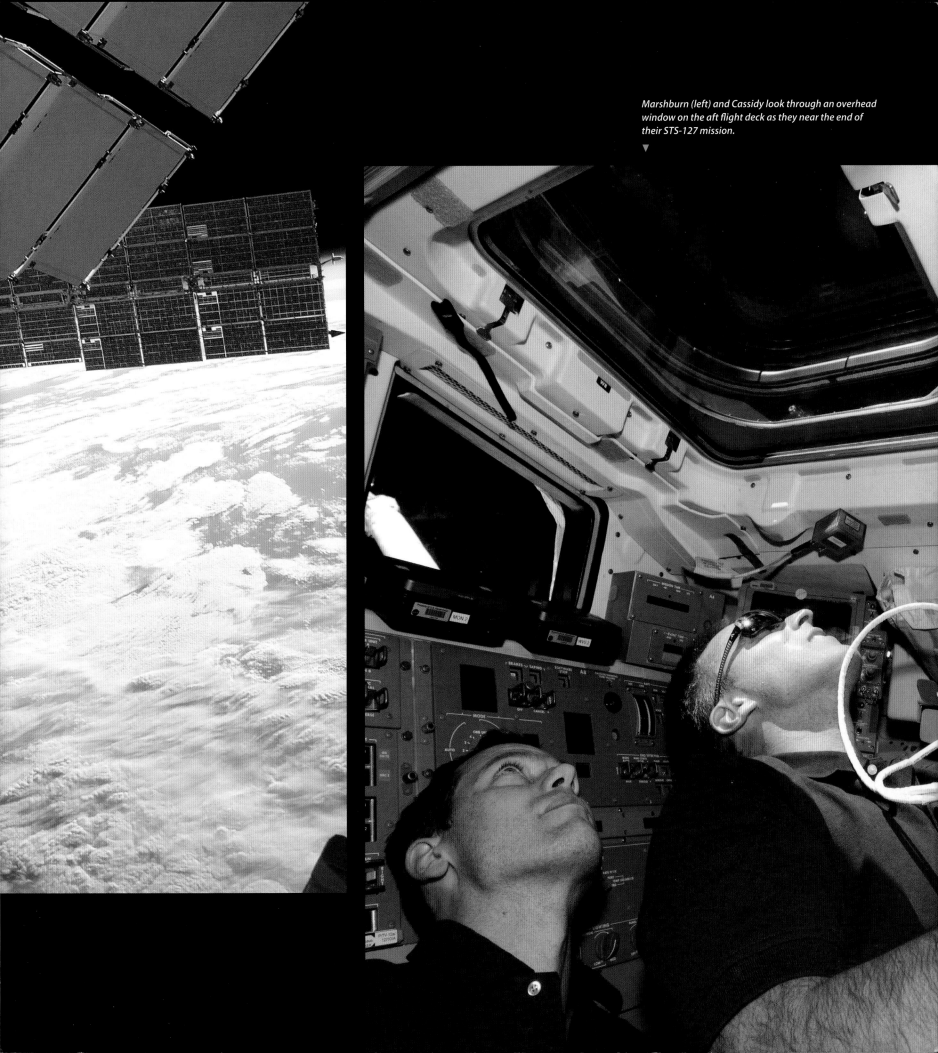

Marshburn (left) and Cassidy look through an overhead window on the aft flight deck as they near the end of their STS-127 mission.

*NASA Administrator (and former shuttle astronaut)
Charles Bolden walks away after welcoming home
STS-127 Endeavour and its crew.*
▼

United States

STS-128 *Discovery*

Launched	August 28, 2009
Landed	September 11, 2009
Commander	Frederick W. Sturckow
Pilot	Kevin Ford

Mission specialists
Patrick Forrester, Christer Fuglesang,
José Hernández, John Olivas

Station crew to orbit
Nicole Stott

Station crew to Earth
Timothy L. Kopra

Milestones
Discovery carried the Leonardo logistics module, containing life-support equipment and science racks. The Lightweight Multi-Purpose Experiment Support Structure Carrier (LMC) was also launched in the payload bay. This was the thirtieth mission of a space shuttle dedicated to the assembly and maintenance of the space station.

A golden glow from sunlight reflected by the station's solar panels, illuminates Olivas (foreground) and Stott as they remove an empty ammonia tank from the truss.

STS-129 *Atlantis*

Launched	November 16, 2009
Landed	November 27, 2009
Commander	Charles Hobaugh
Pilot	Barry Wilmore

Mission specialists
Randy Bresnik, Mike Foreman, Leland Melvin,
Robert Satcher Jr.

Station crew to Earth
Nicole Stott

Milestones
This was the last time that a space shuttle would
transfer crew members to or from the space
station. The eleven-day flight included three space
walks. The payload bay carried two large ExPRESS
logistics carriers holding spare gyroscopes,
nitrogen and ammonia tanks, pumps, and other
items among the complex mass of equipment
demanded to keep the station operational.

▶

*STS-29 stands poised on the pad shortly after the mobile
service structures of the gantry have been rolled back,
clearing the way toward launch.*

In this unusual STS-129 launch photo, the Solid Rocket
Boosters and the External Tank are prominent.

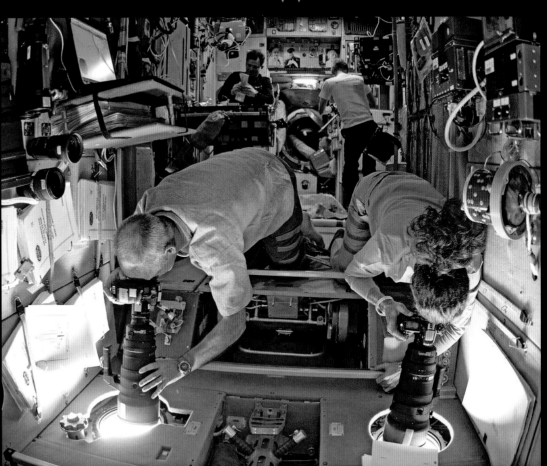

STS-129 astronaut Satcher temporarily occupies the commander's station as he works on one of the various laptop computers augmenting the shuttle's own systems.

On board the station, astronauts Williams (left) and Stott point cameras through separate windows in the Zvezda module as SS-129 Atlantis approaches for docking (below).

▼ ▼

STS-130 *Endeavour*

Launched	February 8, 2010
Landed	February 21, 2010
Commander	George Zamka
Pilot	Terry Virts

Mission specialists
Robert L. Behnken, Kathryn P. Hire,
Nicholas Patrick, Stephen K. Robinson

Milestones

Endeavour delivered a third connecting module, the Tranquility node, to the station, along with a seven-windowed cupola, a spectacular panoramic viewing area to be used as a control room for robotics and docking operations.

NASA celebrated its latest orbital achievements in the same month that a dark shadow fell across its future. Only four flights remained before the shuttle era was expected to end. For the last six years, NASA had been working on a successor vehicle, the Orion capsule, and its stick-shaped rocket, the Ares 1. A large cargo rocket, the Ares V was supposed to be developed in the coming decade, along with a crewed lunar landing vehicle, the Altair. Even as *Endeavour* lifted off, astronauts and ground personnel alike were absorbing the news that the Orion project had been cut back, partly because of budgetary concerns amid a time of economic uncertainty for the global economy.

▲

JAXA astronaut Noguchi has no shortage of camera targets as he tries out the Cupola, just a few hours after its installation by STS-130.

▶
This unique image shows STS-130 Endeavour *approaching the space station. The orange atmospheric layer is the troposphere, where all of the weather and clouds that we typically experience are generated and contained. This layer gives way to the whitish stratosphere, where jet airliners fly, and then into the mesosphere, and the edge of space.*

▶
STS-130 mission specialist Patrick looks out from one of the seven huge windows of the new Cupola.

STS-130's departing view of the nearly completed
International Space Station: our permanently
occupied "island in the sky."

STS-131 *Discovery*

Launched April 5, 2010
Landed April 20, 2010
Commander Alan Poindexter
Pilot James P. Dutton, Jr.

Mission specialists

Clayton Anderson, Dorothy Metcalf-Lindenburger, Rick Mastracchio, Stephanie D. Wilson, Naoko Yamazaki

Milestones

Much attention had been paid to the spectacular view afforded by the cupola window delivered by the previous shuttle mission and attached to the Tranquility node. *Discovery* delivered the Leonardo module, which in turn carried a science rack that fit snugly against a smaller window on the Destiny module. The Window Observational Research Facility, or WORF, named in honor of a popular Klingon character from the *Star Trek* universe, contained multispectral sensors for intensive Earth observation. WORF was the only equipment rack deep enough for the astronauts to work inside, in darkroom conditions suitable for light-sensitive instruments.

Mastracchio (right) and Anderson in the aft section of
the payload bay, in a view highlighting the sheer size and
carrying capability of the shuttle system.

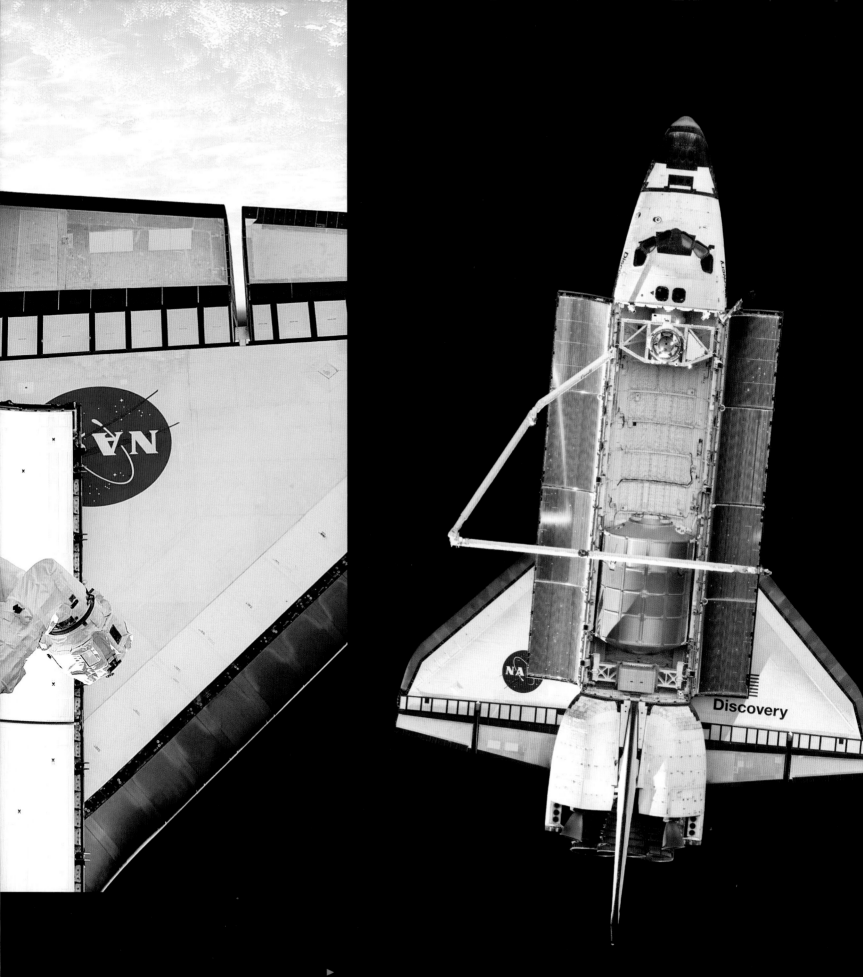

An excellent view of Discovery as it undocks from the
station near the end of the STS-131 mission.

STS-132 *Atlantis*

Launched	May 14, 2010
Landed	May 26, 2010
Commander	Ken Ham
Pilot	Tony Antonelli

Mission specialists
Garrett Reisman, Michael T. Good,
Piers Sellers, Stephen G. Bowen

Milestones

The space shuttle era was drawing to a close. On May 14, *Atlantis* made its last scheduled leap to orbit on mission STS-132, carrying Rassvet ("dawn"), the second of two Russian mini research modules, plus spare batteries for the station's main truss and a new communications dish to improve the rate at which data could be swapped between the station and the ground. In the twenty-five years since its debut in October 1985, *Atlantis* accumulated more than three hundred days in space, making thirty-two return trips to orbit, including seven to the Mir space station in the 1990s and eleven to the International Space Station thereafter. It flew five classified military missions, launched two planetary probes (Magellan to Venus and Galileo to Jupiter), deployed the Compton gamma ray observatory and gave the Hubble Space Telescope its final overhaul. By any measure, *Atlantis* was a wonderful spacecraft.

▶

Ham in the commander's seat for STS-132, wearing the pressure suit and parachute pack as protection against accidents. As history has shown, these garments, though perfectly well-made, were not sufficient, on their own, to guard crew members against the hazards of launch or re-entry failures.

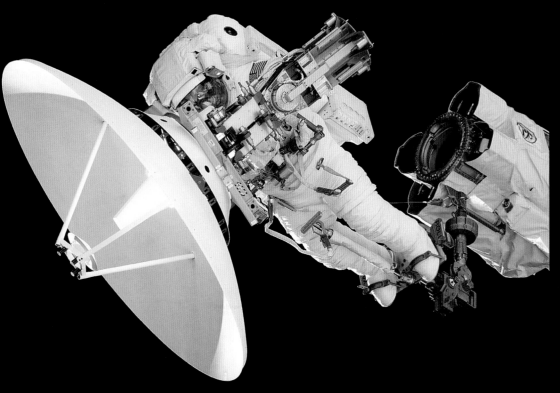

Reisman and Bowen (out of frame) installed a second antenna for high-speed transmissions and added a two-armed extension to the station's robotic arm.

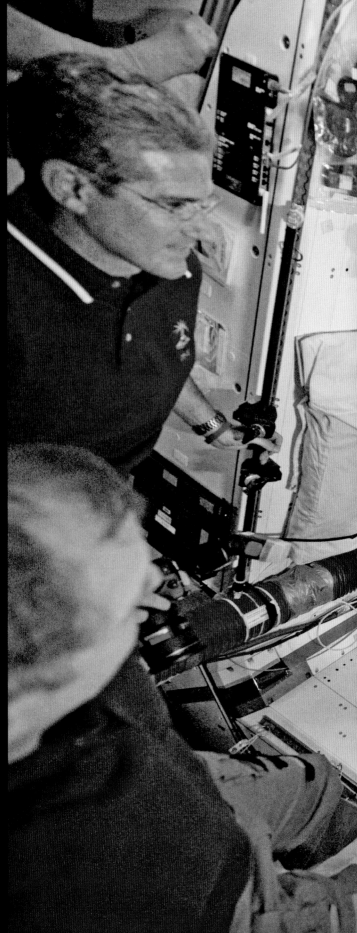

▶

STS-132 commander Ham (center frame) and Russian cosmonaut Oleg Kotov, space station commander at that time, shake hands during a farewell ceremony in the Harmony node, shortly before Atlantis leaves for home.

In this wide-angle view, Reisman is surrounded by windows and computer screens in the Cupola, in a scene that brings science fiction predictions to life.

This beautiful shot of the orbiter's nose during the STS-132 mission was a sharp reminder to NASA's archivists that not one photo has ever been taken of a shuttle docked to the ISS in which the entire orbiter is visible.

STS-133 *Discovery*

Launched February 24, 2011
Landed March 9, 2011
Commander Steve Lindsey
Pilot Eric Boe

Mission specialists
Alvin Drew, Nicole Stott, Steve Bowen,
Michael Barratt

Milestones
This mission delivered a Multi-Purpose Logistics
Module and an external stowage platform, both
for permanent attachment to the Space Station.
Bowen and Drew performed two spacewalks for
maintenance work and installation of new
components. Among the cargo was Robonaut, a
human-shaped robotic torso, head, and pair of
arms and hands, that will render assistance to
astronauts during future spacewalks. Robonaut is
controlled by human operators inside the space
station and can also be operated remotely by
ground controllers on Earth.

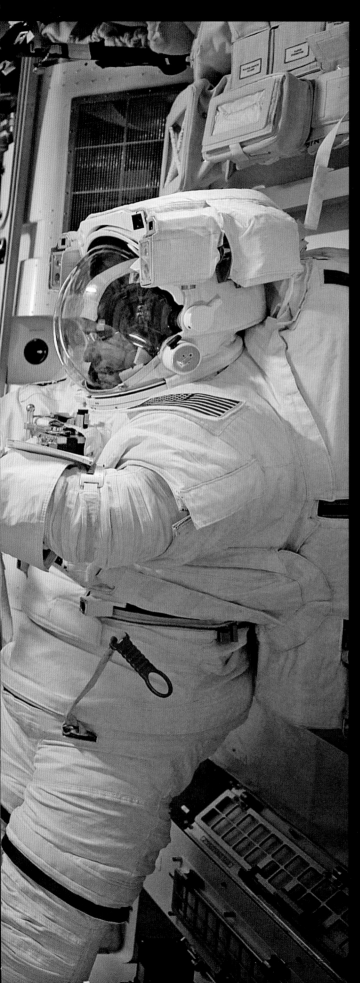

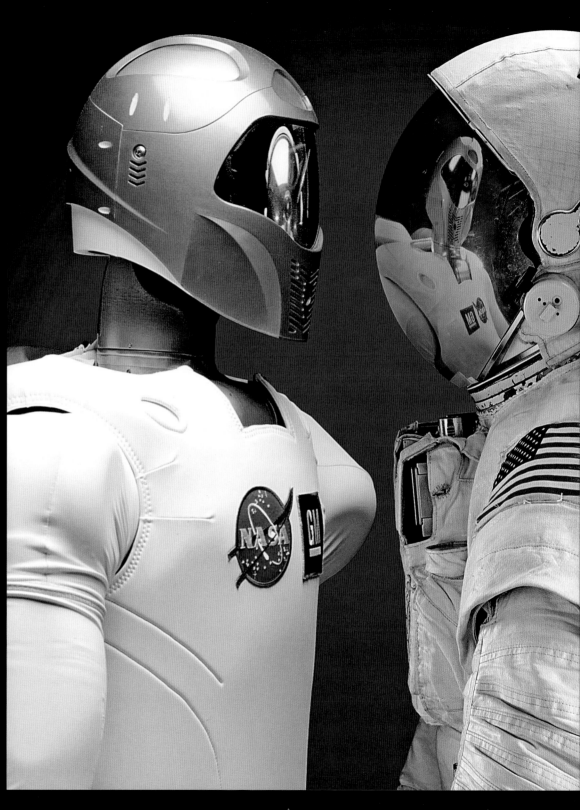

Robonaut confronts a human co-worker in a staged press appearance prior to the launch of STS-133.

STS-133 astronauts Bowen (left), Barratt (center), and Drew in the Quest airlock of the International Space Station as they prepare for the start of the mission's first spacewalk.

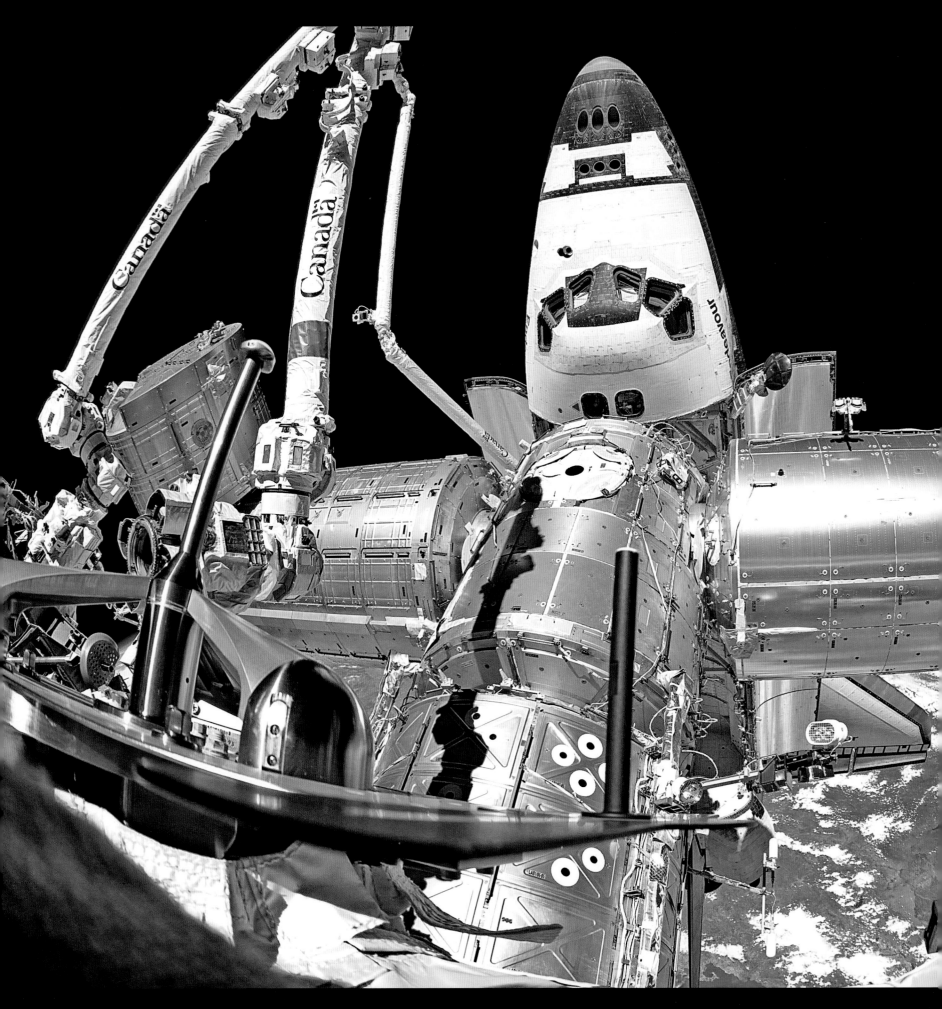

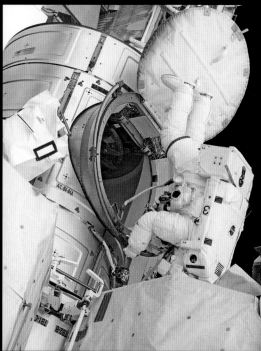

▲ STS-134 astronaut Feustel clambers around the open hatch of the Quest airlock.

STS-134 *Endeavour*

Launched	May 16, 2011
Landed	June 1, 2011
Commander	Mark E. Kelly
Pilot	Gregory H. Johnson

Mission specialists
Michael Fincke, Greg Chamitoff, Andrew Feustel, Roberto Vittori

Milestones
The star payload for *Endeavour's* last flight was a particle physics instrument, the Alpha Magnetic Spectrometer (AMS), for installation on the exterior of the International Space Station. The $2 billion AMS will help researchers study the formation of the universe and find evidence of dark matter, the unknown material that makes up the bulk of the universe's mass. The instrument will also seek out cosmic sources of antimatter particles. *Endeavour* also carried spare parts for the station, including two communications antennas and components for the station's robotic arms. Commander Kelly was informed, mid-flight, that a delicate surgical operation on his wife, Arizona Congresswoman Gabrielle Giffords, was successful. Giffords is making a remarkable recovery after being shot in the head in January 2011 by a deranged gunman. (Six people were killed during this incident.) A unique media highlight of the STS-134 mission was the in-flight photography conducted by the crew of a departing Soyuz capsule heading for home. On board were homeward-bound station personnel Dimitri Kondratyev, Paolo Nespoli, and Cady Coleman. Nesploli, a keen photographer, took unprecedented shots of *Endeavour* docked to the station, as seen from a distance.

◀ A portion of the space station and the docked shuttle Endeavour *is featured in this image, photographed with a fish-eye lens attached to an electronic still camera.*

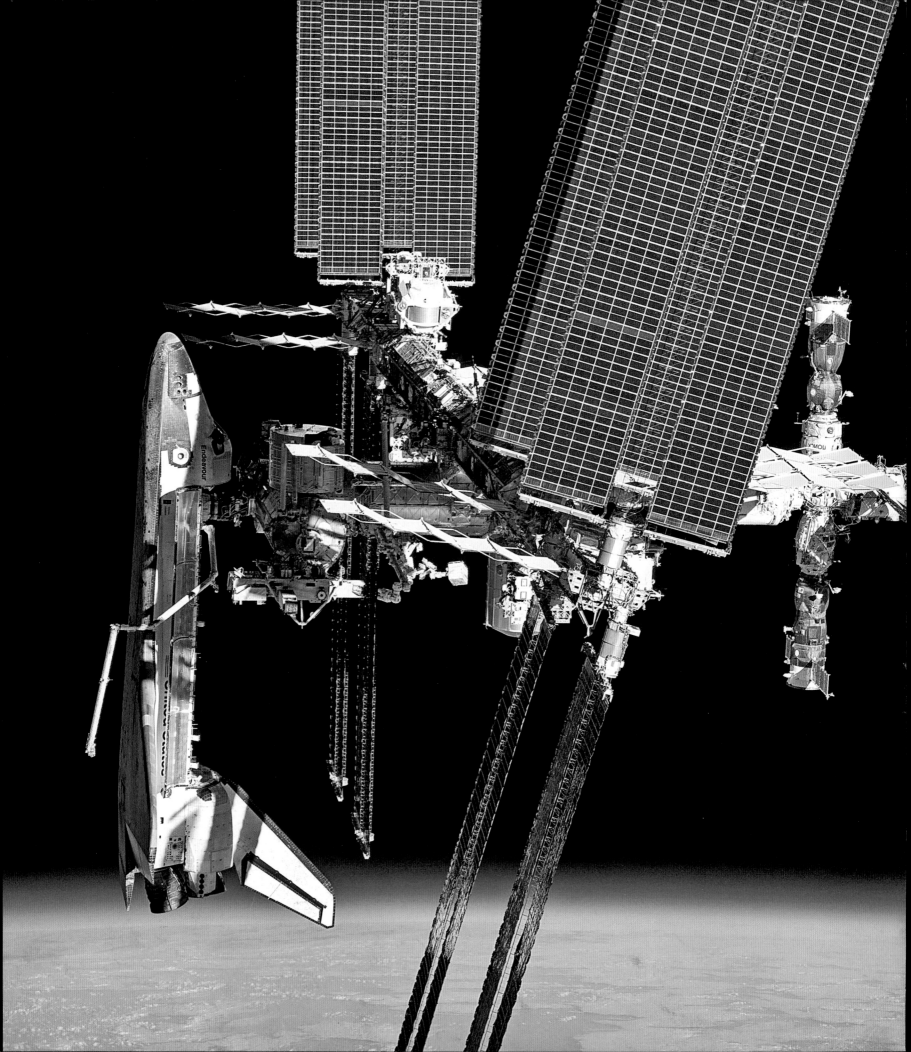

The docked STS-134 Endeavour is backdropped by a night time view of Earth and a starry sky, in a rare long exposure photograph taken through a window on the space station. Most images of spacecraft in orbit do not reveal the stars, because typical exposure times are too brief to capture their dim light.

An unprecedented photo of Endeavour docked to the space station, and seen from afar, as taken by ESA astronaut Paolo Nespoli while aboard a departing Soyuz spacecraft.

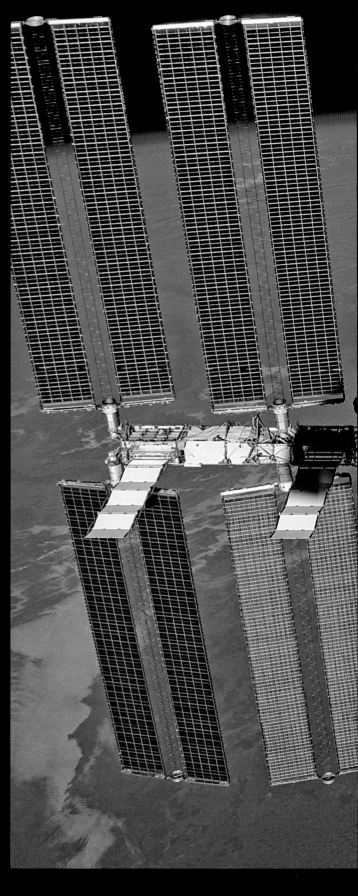

▲
Commander Kelly has every reason to look relieved at the end of his successful flight and after hearing the good news about Gabrielle's medical operation.

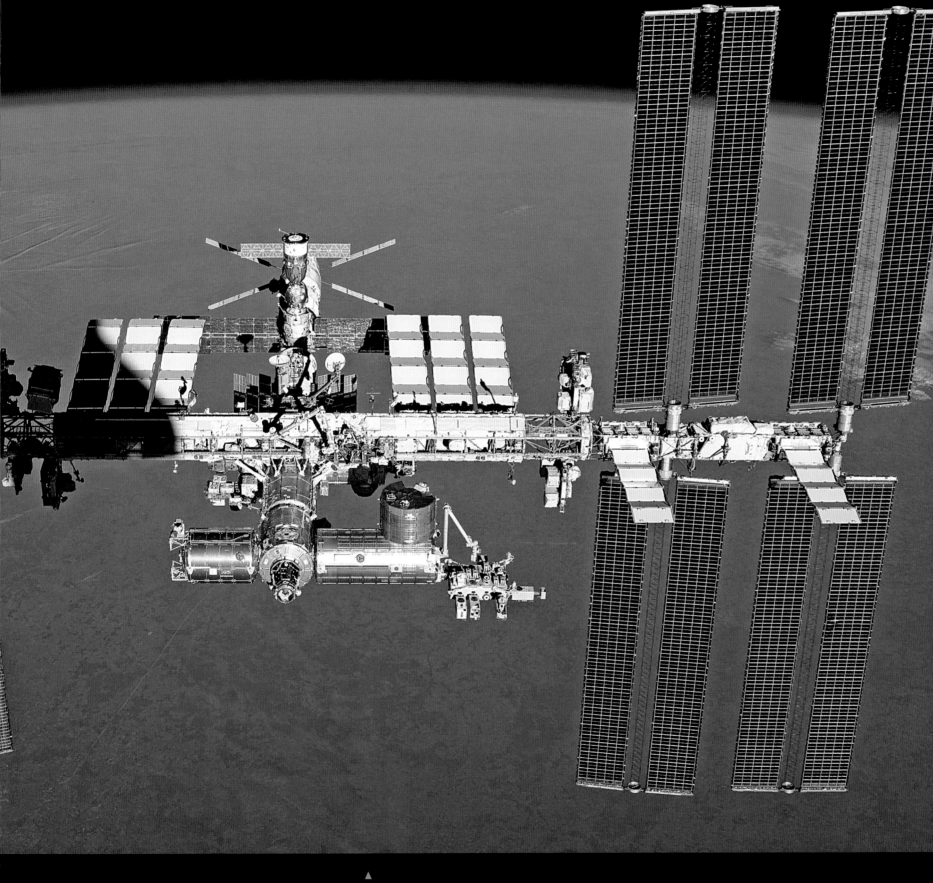

▲
The fully assembled International Space Station, seen by the crew of STS-134 as Endeavour *moves away at the conclusion of its successful mission.*

STS-135 *Atlantis*

Scheduled   July 2011
Commander   Christopher Ferguson
Pilot   Douglas Hurley

Mission specialists
Sandra Magnus, Rex Walheim

Milestones

This is the last flight for *Atlantis*, and indeed for
the entire space shuttle fleet. The four-person
crew complement ensures that the crew can
remain aboard the space station if *Atlantis* suffers
any significant damage to its heat shielding. A
Soyuz capsule would then be used to bring the
STS-135 crew back to Earth. The mission will
deliver supplies to the station, using a Multi-
Purpose Logistics Module. *Atlantis* will also carry
special hardware for the station's robotic arms,
aimed at simulating the remote-controlled
refueling of satellites in space.

▶

*STS-135 crew members pose for a group photo, while
Atlantis sits in the Orbiter Processing Facility in April 2011,
being made ready for the last shuttle flight. The crew are
surrounded by NASA staffers and contractor personnel
representing just a small fraction of the team that has
serviced and maintained the shuttle fleet.*

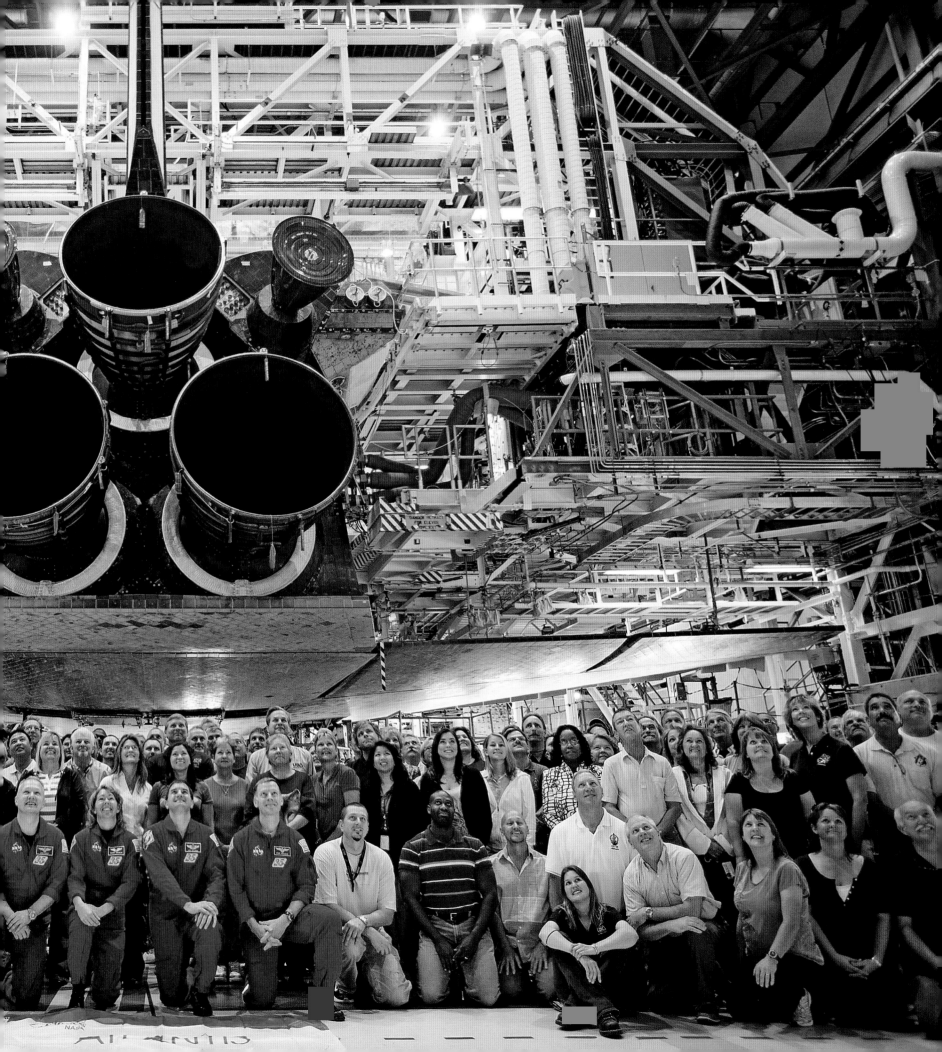

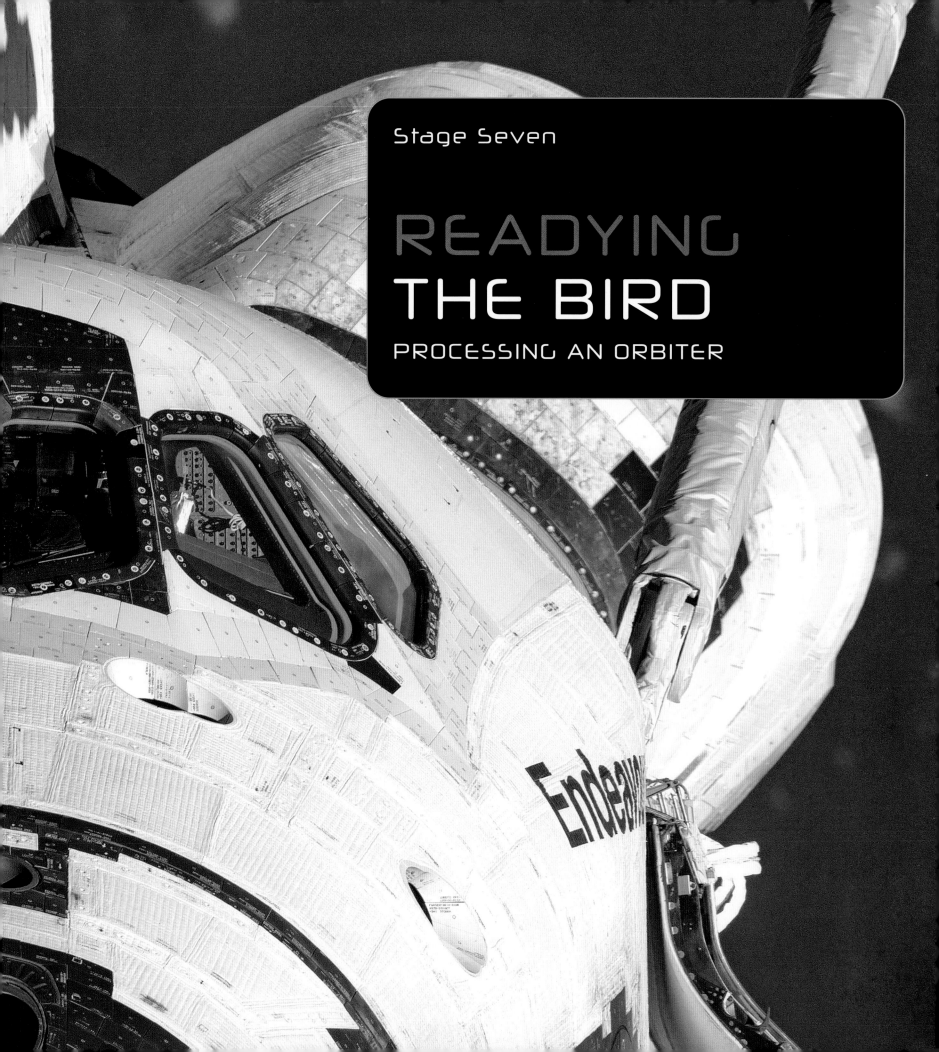

READYING
THE BIRD
PROCESSING AN ORBITER

READYING THE BIRD
PROCESSING AN ORBITER

Astronauts were the main human focus of orbital missions, but their success was reliant upon the shuttle processing crews and pad technicians: the thousands of people behind the scenes who made the vehicle fly.

IMMEDIATELY ON ITS RETURN from space, a shuttle orbiter is "safed." All propellant remnants and other dangerous residues are piped away from the tanks, engine pipework and thruster systems.

If an orbiter has touched down at Edwards Air Force Base in Cailifornia, it will be towed into the Mate-Demate Device for installation atop the 747 Shuttle Carrier Aircraft, ready for the flight back to the Kennedy Space Center in Florida.

Back at Kennedy, the orbiter begins its complex refurbishment in an Orbiter Processing Facility (OPF), one of three 95-foot-tall hangars located near the famous Vehicle Assembly Building (VAB) at the Kennedy Space Center. The relevant mission payloads are installed, and a complete check of computers, avionics, flap and aileron actuators, and other powered systems is conducted. Meanwhile the huge External Tank (ET) arrives by barge from the assembly plant in Michoud, Louisiana, and is unloaded and moved to the VAB, where it is pulled into vertical position by cranes.

Within one of four huge bays within the VAB, the lower skirts of two SRBs rest on a rectangular mobile launch platform, a two-story steel structure 25 feet high, 160 feet long, and 135 feet wide, constructed from massive six-inch-thick steel plates. Each SRB is secured to the platform by four bolts that are instantly severed at liftoff by small explosives.

The twin Solid Rocket Boosters (SRBs) and the ET are mated in the largest VAB structure, a cavernous hangar more than 500 feet tall, originally constructed for the stacking of Saturn V Moon rockets, and divided into four areas known, appropriately enough, as High Bays. The orbiter is towed on a special flatbed trailer from the

OPF and into the VAB, where it is grappled by a rigid lifting frame attached to wires and pulleys suspended from the upper levels of the building. The orbiter is lifted from a horizontal to a vertical posture, then carefully oriented so that it can be lifted upward yet farther, so that it comes alongside the external tank.

Once all the launch components are mated together, the entire stack is carried to Pad 39A or 39B by the crawler-transporter, moving on four double-tracked caterpillars. The crawler maneuvers underneath the launch platform, then takes its entire weight and brings it out of the VAB for the slow and careful move to the launch pad: a journey that spans less than four miles but takes at least five hours to complete.

Assuming all is well, at T-0 the explosive bolts securing the SRBs to the launch platform are severed. The SRBs are ignited by pyrotechnic wires sparking along the entire length of the propellant lining. The instant they come to life, the shuttle lifts off the pad and begins its ascent to orbit.

The very instant that a crewed launch vehicle leaves the pad at Kennedy, it is deemed to be in flight and conducting its mission. At this moment, all ground control functions are taken over by NASA's Mission and Operations Control Center (MOCR) at the Johnson Space Center (JSC) near Houston, Texas.

Smoke and fire

The vast cloud of smoke attending liftoff consists mainly of steam from the water sprayed onto the pad, which is instantly superheated by the combined fury of five rocket exhausts blasting down on it. By now, the main engines are consuming their liquid hydrogen and

This close-up shot shows the "safing" pipes removing dangerous residues from the tanks and pipework of an orbiter's propulsion and thruster systems.

star-shaped cross-section, running the entire length of the solid fuel lining.

Seven seconds into its flight, the shuttle rolls on its axis by 120 degrees, adopting the correct aerodynamic position for the main ascent. This "heads-down" posture ensures that SRB and ET separations will occur smoothly.

By 50 seconds, the shuttle has broken through the speed of sound. At 2 minutes and 12 seconds, the vehicle is at an altitude of at 28 miles (45 kilometers) and flying at 4.5 times the speed of sound. At this point the SRBs have exhausted their fuel and are discarded by small explosive bolts that push them clear of the ET. The SRBs descend by parachute and are subsequently recovered from the Atlantic by special tug boats.

At 8 minutes and 30 seconds, the three main engines are shut down, and 20 seconds later, the empty, nonrecoverable ET separates and begins to fall back toward the Earth—or more specifically, the Atlantic and a watery grave for whatever wreckage is left after a fiery atmospheric descent and subsequent high-speed impact with the water.

The two engines of the Orbital Maneuvering System (OMS), fueled purely from within the orbiter's reserves, now fire to push the vehicle into a low, elliptical (egg-shaped) orbit. The OMS engines burn monomethyl hydrazine fuel and nitrogen tetroxide oxidizer, volatile chemicals that ignite spontaneously on contact without the need for a spark. These fuels are pushed through the pipework by harmless and chemically inert helium gas. Pressurized nitrogen gas is used to open and close the engine valves.

Around 45 minutes later, after a half-circuit of the Earth has been completed, the OMS fires once more to put the vehicle into its final circular orbit.

The forty-four small rocket engines of the Reaction Control System (RCS) enable the orbiter to make pitch, roll, and yaw movements relative to the Earth's horizon and to adjust its velocity with extreme precision, especially when making a rendezvous approach. The RCS thrusters, which are all fueled and pressurized in a similar manner to the OMS engines, are grouped in three modules: one in the orbiter's nose and one in each of the twin OMS pods.

At the end of a mission, the OMS engines will be fired once more, but this time against the orbiter's west-east direction of travel, to decelerate the vehicle ahead of re-entry, which is fine-tuned with small thrusters. From here on, the craft is unpowered. It glides to touchdown purely under the influence of its wing control surfaces: the tail fin rudders and airbrakes, and the wing ailerons.

oxygen from the External Tank (ET) at a rate that would suck a family swimming pool dry in under half a minute.

A smaller cloud of yellowish smoke results from the SRB's combustion of aluminum powder fuel, ammonium perchlorate oxidizer, and iron oxide catalyst, bound together in a sticky paste. The combustion chamber for an SRB is essentially a long, hollow groove, with a

A recently landed orbiter is towed toward a processing hangar for post-flight refurbishment.

Orbiters landing at Edwards Air Force Base in California are attached to the Shuttle Carrier Aircraft for return to the Kennedy launch complex in Florida. The gantry that accomplishes this task is called the Mate-Demate Device.

A rear view of an orbiter atop the Shuttle Carrier Aircraft at the start of a return trip to the Kennedy Space Center.

Inside one of three Orbiter Processing Facility hangars at Kennedy, work begins on refreshing an orbiter's systems in readiness for its next flight.

The External Tanks are assembled at NASA's Michoud Assembly Facility in New Orleans, Louisiana: a veteran building where the gigantic first stages for Saturn V rocket stages were once assembled.

Once completed, an ET is towed on a barge across the Gulf of Mexico, around Florida, and up to the Kennedy Space Center.

After unloading from its barge, an External Tank is pulled into the VAB to be stacked with the Solid Rocket Boosters.

The Solid Rocket Boosters arrive at the VAB in short segments that will be mated together via O-ring joints.

The O-ring interfaces of a Solid Rocket Booster are fitted together with extreme precision, in a bid to eliminate the slightest risk of leakage from hot ignition gases.

Technicians standing at the base of the SRB and ET stack show the impressive scale of the shuttle system.

Refurbished orbiters are reversed out of the relevant processing hangar on a special flatbed trailer. Their next stop will be the Vertical Assembly Buidling (VAB).

A closer view of the flatbed trailer, as an orbiter arrives at the VAB to be mated with the other components of the launch stack.

A complex series of lifts, twists, and reorientations is required as the orbiter is lifted from the floor of the VAB and slowly positioned belly-first against the awaiting External Tank.

Despite the seeming immensity of the shuttle hardware, the VAB building makes even the largest components look small. The building was originally intended to service several Saturn rockets simultaneously. That dream was seldom fulfilled. In a sense, the VAB has never had the chance to work at its fullest capacity.

A fully assembled shuttle launch stack emerges from the VAB on the crawler-transporter platform, and makes its way slowly toward the launch pad.

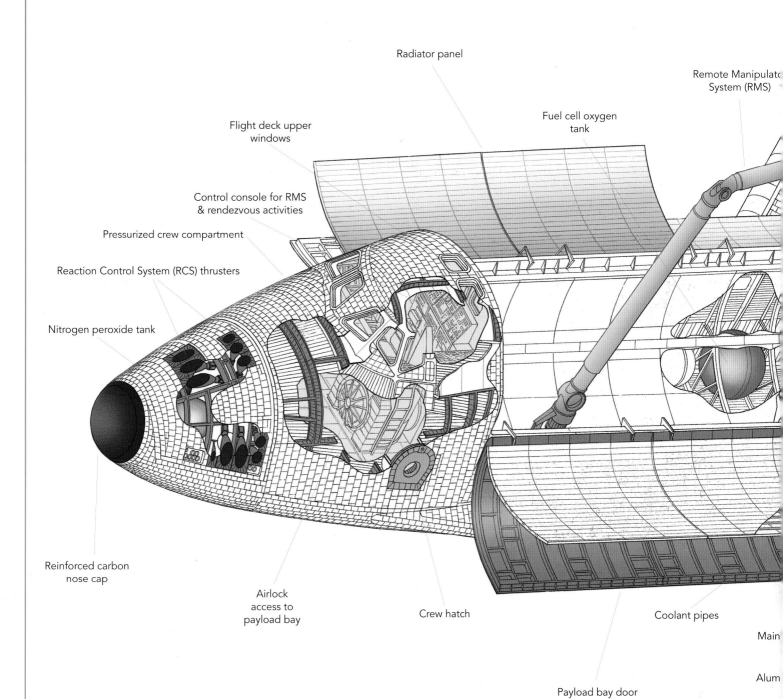

Radiator panel

Remote Manipulato
System (RMS)

Flight deck upper
windows

Fuel cell oxygen
tank

Control console for RMS
& rendezvous activities

Pressurized crew compartment

Reaction Control System (RCS) thrusters

Nitrogen peroxide tank

Reinforced carbon
nose cap

Airlock
access to
payload bay

Crew hatch

Coolant pipes

Main

Alum

Payload bay door

SPACE SHUTTLE ORBITER

CUTAWAY VIEW OF COLUMBIA AS CONFIGURED IN 1981

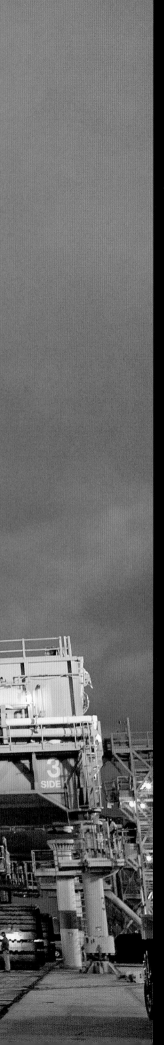

COUNTDOWN & IGNITION

A countdown begins at about T-43 hours ("T" being the moment of liftoff, and the dash representing "minus"). All launch preparations are conducted from the Launch Control Center (LCC). This seemingly humble building sits adjacent to the VAB. From here, all shuttle missions are monitored and controlled up to the point of ignition.

Just over twenty-four hours before liftoff, the orbiter is primed with its internal load of propellants and fuels for its Reaction Control System (RCS) maneuvering thrusters and twin Orbital Maneuvering System (OMS) engines. Hydrogen and oxygen gases are also loaded into the fuel cells that generate electrical power for on-board systems.

Loading of liquid hydrogen and liquid oxygen into the external tank begins at T-6 hours and takes three hours. By this point, only specialized personnel are permitted on the pad, such as the final inspection team that conducts a walking inspection of the vehicle.

Another team prepares the white room, the mobile box that swings against the orbiter's hatch to enable crew members to get into the vehicle, two hours prior to launch, along with a team of technicians who secure the astronauts into their couches, and check that the flight deck is properly configured.

At T-20 minutes the launch director solicits a "go" from all ground controllers, and the countdown proceeds to T-9 minutes, from which point the final sequence is under computer control, albeit monitored closely by humans.

Nine seconds before ignition, the orbiter's three main engines undergo a gimbal test, ensuring their ability to swivel ("gimbal") for steering, under command of the on-board guidance computers. A pyrotechnic device just beneath the main engines is fired up to make sure that all escaping engine gases are properly ignited. A spray of water, gushing at the rate of 900,000 gallons per minute, swamps the lower regions of the launch pad to suppress the thermal shock and damp down the violent sound vibrations that are about to be unleashed.

Six seconds prior to liftoff, fuel is surging into the main engines, and they ignite. The tip of the launch stack shifts about forty inches from the vertical, but for the next five seconds, the vehicle stays attached to the pad. The Solid Rocket Boosters (SRBs) are not lit until confirmation is received that all three main engines are running correctly. Once lit, the SRBs cannot be switched off in the event of any problems. They can only be allowed to fire if all other systems in the shuttle stack are functioning flawlessly.

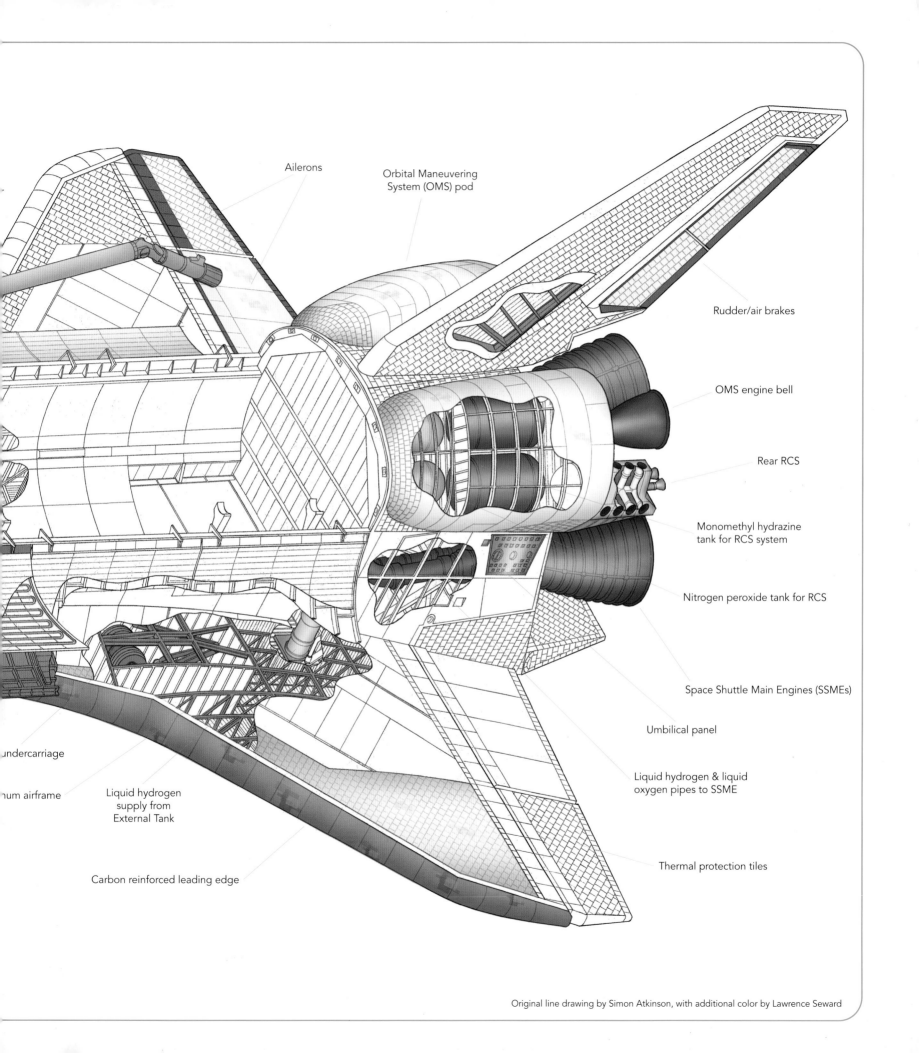

Ailerons

Orbital Maneuvering
System (OMS) pod

Rudder/air brakes

OMS engine bell

Rear RCS

Monomethyl hydrazine
tank for RCS system

Nitrogen peroxide tank for RCS

Space Shuttle Main Engines (SSMEs)

Umbilical panel

Liquid hydrogen & liquid
oxygen pipes to SSME

undercarriage

num airframe

Liquid hydrogen
supply from
External Tank

Carbon reinforced leading edge

Thermal protection tiles

Original line drawing by Simon Atkinson, with additional color by Lawrence Seward

SHUTTLE FLIGHT LOG
COMPLETE MISSION SUMMARY

Sequence	Launch Date	STS number	Orbiter	Duration	Landing site	Mission highlights	Page
01	April 12, 1981	STS-1	*Columbia*	2 days, 6 hours	Edwards	First flight to orbit of the STS system	024
02	Nov. 12, 1981	STS-2	*Columbia*	2 days, 6 hours	Edwards	First reuse of an orbiter; first test of Canadarm	033
03	March 22, 1982	STS-3	*Columbia*	8 days, 0 hours	White Sands	Unique landing in New Mexico	034
04	June 27, 1982	STS-4	*Columbia*	7 days, 1 hour	Edwards	Department of Defense (DoD) payload	035
05	Nov. 11, 1982	STS-5	*Columbia*	5 days, 2 hours	Edwards	Multiple comsat launches; first EVA cancelled	036
06	April 4, 1983	STS-6	*Challenger*	5 days, 0 hours	Edwards	Launch of TDRS; first EVA; first flight for *Challenger*	037
07	June 18, 1983	STS-7	*Challenger*	6 days, 2 hours	Edwards	Sally Ride, first US woman in space	038
08	Aug. 30, 1983	STS-8	*Challenger*	6 days, 1 hour	Edwards	First African-American in space, Guion Bluford	040
09	Nov. 28, 1983	STS-9	*Columbia*	10 days, 7 hours	Edwards	First flight of the European-built Spacelab	041
10	Feb. 3, 1984	STS-41B	*Challenger*	7 days, 23 hours	Kennedy	Bruce McCandless conducts untethered EVA	042
11	April 6, 1984	STS-41C	*Challenger*	6 days, 23 hours	Edwards	First satellite rescue of Solar Max; LDEF deployment	048
12	Aug. 30, 1984	STS-41D	*Discovery*	6 days, 0 hours	Edwards	First flight for *Discovery*; multiple comsat launches	050
13	Oct. 5, 1984	STS-41G	*Challenger*	8 days, 5 hours	Kennedy	Kathryn Sullivan, first US woman to conduct an EVA	050
14	Nov. 8, 1984	STS-51A	*Discovery*	7 days, 23 hours	Kennedy	Retrieval of Palapa B2 and Westar VI satellites	051
15	Jan. 24, 1985	STS-51C	*Discovery*	3 days, 1 hours	Kennedy	Classified DoD mission	053
16	April 12, 1985	STS-51D	*Discovery*	6 days, 23 hours	Kennedy	US senator Jake Garn on board; comsats launched	053
17	April 29, 1985	STS-51B	*Challenger*	7 days, 0 hours	Edwards	Second Spacelab mission	054
18	June 17, 1985	STS-51G	*Discovery*	7 days, 1 hour	Edwards	Multiple comsat launches	054
19	July 29, 1985	STS-51F	*Challenger*	7 days, 22 hours	Edwards	Third Spacelab mission	055
20	Aug. 27, 1985	STS-51I	*Discovery*	7 days, 2 hours	Edwards	Rescue of Syncom F3 (Leasat-3) satellite	056
21	Oct. 3, 1985	STS-51J	*Atlantis*	4 days, 1 hour	Edwards	First flight for *Atlantis* on classified DoD mission	056
22	Oct. 30, 1985	STS-61A	*Challenger*	7 days, 0 hours	Edwards	*Challenger*'s last successful flight, with Spacelab	056
23	Nov. 26, 1985	STS-61B	*Atlantis*	6 days, 21 hours	Edwards	Multiple comsat launches; ACCESS experiment	057

Sequence	Launch Date	STS number	Orbiter	Duration	Landing site	Mission highlights	Page
24	Jan. 12, 1986	STS-61C	*Columbia*	6 days, 2 hours	Edwards	Congressman Bill Nelson aboard; comsat launch	059
25	Jan. 28, 1986	STS-51L	*Challenger*	73 seconds	n/a	Loss of vehicle and crew during ascent	062
26	Sept. 29, 1988	STS26	*Discovery*	4 days, 1 hour	Edwards	Reverification flight; deployment of TDRS	074
27	Dec. 2, 1988	STS-27	*Atlantis*	4 days, 9 hours	Edwards	DoD mission, including launch of Lacrosse 1	074
28	March 13, 1989	STS-29	*Discovery*	4 days, 23 hours	Edwards	Space station radiator test; IMAX camera aboard	075
29	May 4, 1989	STS-30	*Atlantis*	4 days, 0 hours	Edwards	Launch of Magellan probe to Venus	076
30	Aug. 8, 1989	STS-28	*Columbia*	5 days, 1 hour	Edwards	Classified DoD mission with a military comsat	077
31	Oct. 18, 1989	STS-34	*Atlantis*	4 days, 23 hours	Edwards	Launch of Galileo Jupiter probe; IMAX aboard	077
32	Nov. 22, 1989	STS-33	*Discovery*	5 days, 0 hours	Edwards	Classified DoD Mission with reconnaissance satellite	078
33	Jan. 9, 1990	STS-32	*Columbia*	10 days, 21 hours	Edwards	Launch of Syncom; retrieval of LDEF	079
34	Feb. 28, 1990	STS-36	*Atlantis*	4 days, 10 hours	Edwards	DoD mission, with Misty reconnaissance satellite	080
35	April 24, 1990	STS-31	*Discovery*	5 days, 1 hour	Edwards	Launch of the Hubble Space Telescope	080
36	Oct. 6, 1990	STS-41	*Discovery*	4 days, 2 hours	Edwards	Ulysses solar probe launch	082
37	Nov. 15, 1990	STS-38	*Atlantis*	4 days, 21 hours	Kennedy	DoD mission to deploy reconnaissance satellite	082
38	Dec. 2, 1990	STS-35	*Columbia*	8 days, 23 hours	Edwards	ASTRO-1 astronomical telescopes on board	083
39	April 5, 1991	STS-37	*Atlantis*	5 days, 23 hours	Edwards	Launch of Compton Gamma Ray Observatory	083
40	April 28, 1991	STS-39	*Discovery*	8 days, 7 hours	Kennedy	Unclassified DoD mission with military experiments	084
41	June 5, 1991	STS-40	*Columbia*	9 days, 2 hours	Edwards	Spacelab mission	085
42	Aug. 21, 1991	STS-43	*Atlantis*	8 days, 2 hours	Kennedy	Tracking & Data Relay Satellite launch	086
43	Sept. 12, 1991	STS-48	*Discovery*	5 days, 8 hours	Edwards	Upper Atmosphere Research Satellite launch	087
44	Nov. 24, 1991	STS-44	*Atlantis*	6 days, 22 hours	Edwards	Defense Support Program (DSP) Satellite	088
45	Jan. 22, 1992	STS-42	*Discovery*	8 days, 1 hour	Edwards	Spacelab mission	088
46	March 24, 1992	STS-45	*Atlantis*	8 days, 22 hours	Kennedy	ATLAS-1 science platform	089
47	May 7, 1992	STS-49	*Endeavour*	8 days, 21 hours	Edwards	Intelsat VI repair; space station truss test	090
48	June 15, 1992	STS-50	*Columbia*	13 days, 19 hours	Kennedy	Spacelab mission	092
49	July 31, 1992	STS-46	*Atlantis*	7 days, 23 hours	Kennedy	European Retrievable Carrier (EURECA) deployed	093
50	Sept. 12, 1992	STS-47	*Endeavour*	7 days, 22 hours	Kennedy	Spacelab joint mission with Japan's space agency	093
51	Oct. 22, 1992	STS-52	*Columbia*	9 days, 20 hours	Kennedy	Laser Geodynamics Satellite II (LAGEOS-II)	094

Sequence	Launch Date	STS number	Orbiter	Duration	Landing site	Mission highlights	Page
52	Dec. 2, 1992	STS-53	Discovery	7 days, 7 hours	Edwards	Final classified DoD mission, with satellite deployment	095
53	Jan. 13, 1993	STS-54	Endeavour	5 days, 23 hours	Kennedy	Tracking & Data Relay Satellite (TDRS) deployment	095
54	April 8, 1993	STS-56	Discovery	9 days, 6 hours	Kennedy	ATLAS-2 science platform	096
55	April 26, 1993	STS-55	Columbia	9 days, 23 hours	Edwards	Spacelab mission	096
56	June 21, 1993	STS-57	Endeavour	9 days, 23 hours	Kennedy	SpaceHab logistics module; EURECA retrieved	098
57	Sept. 12, 1993	STS-51	Discovery	9 days, 20 hours	Kennedy	Advanced Communications Technology Satellite	100
58	Oct. 18, 1993	STS-58	Columbia	14 days, 0 hours	Edwards	Spacelab mission	101
59	Dec. 2, 1993	STS-61	Endeavour	10 days, 19 hours	Kennedy	Hubble Space Telescope servicing mission	102
60	Feb. 3, 1994	STS-60	Discovery	7 days, 6 hours	Kennedy	SpaceHab, with Wake Shield Facility	107
61	March 4, 1994	STS-62	Columbia	13 days, 23 hours	Kennedy	Microgravity experiments	107
62	April 9, 1994	STS-59	Endeavour	11 days, 5 hours	Edwards	Earth topographical survey via radar	108
63	July 8, 1994	STS-65	Columbia	14 days, 17 hours	Kennedy	Spacelab mission	109
64	Sept. 9, 1994	STS-64	Discovery	10 days, 22 hours	Edwards	Multiple science experiments	110
65	Sept. 30, 1994	STS-68	Endeavour	11 days, 5 hours	Edwards	Earth topographical survey via radar	112
66	Nov. 3, 1994	STS-66	Atlantis	10 days, 22 hours	Edwards	ATLAS-3 science platform	113
67	Feb. 3, 1995	STS-63	Discovery	8 days, 6 hours	Kennedy	Mir close approach; SpaceHab; IMAX camera	122
68	March 2, 1995	STS-67	Endeavour	16 days, 15 hours	Edwards	Spacelab mission, including ultraviolet astronomy	123
69	June 27, 1995	STS-71	Atlantis	9 days, 19 hours	Kennedy	First Shuttle-Mir docking	123
70	July 13, 1995	STS-70	Discovery	8 days, 22 hours	Kennedy	Tracking & Data Relay Satellite (TDRS) deployment	127
71	Sept. 7, 1995	STS-69	Endeavour	10 days, 20 hours	Kennedy	Wake Shield Facility; solar physics instruments	128
72	Oct. 20, 1995	STS-73	Columbia	15 days, 21 hours	Kennedy	Spacelab mission	130
73	Nov. 12, 1995	STS-74	Atlantis	8 days, 4 hours	Kennedy	Shuttle-Mir docking module delivered; IMAX aboard	131
74	Jan. 11, 1996	STS-72	Endeavour	8 days, 22 hours	Kennedy	Japan's Space Flyer Unit (SFU) retrieved	132
75	Feb. 22, 1996	STS-75	Columbia	15 days, 17 hours	Kennedy	Tethered Satellite system	132
76	March 22, 1996	STS-76	Atlantis	9 days, 5 hours	Edwards	Shuttle-Mir docking	132
77	May 19, 1996	STS-77	Endeavour	10 days, 0 hours	Kennedy	SpaceHab commercial space product experiments	136
78	June 20, 1996	STS-78	Columbia	16 days, 21 hours	Kennedy	Spacelab mission	136
79	Sept. 16, 1996	STS-79	Atlantis	10 days, 3 hours	Kennedy	Shuttle-Mir docking	137

Sequence	Launch Date	STS number	Orbiter	Duration	Landing site	Mission highlights	Page
80	Nov. 19, 1996	STS-80	*Columbia*	17 days, 15 hours	Kennedy	Longest mission, for astronomy & Wake Shield	138
81	Jan. 12, 1997	STS-81	*Atlantis*	10 days, 4 hours	Kennedy	Shuttle-Mir docking	138
82	Feb. 11, 1997	STS-82	*Discovery*	9 days, 23 hours	Kennedy	Hubble Space Telescope servicing mission	139
83	April 4, 1997	STS-83	*Columbia*	3 days, 23 hours	Kennedy	Microgravity Science, cut short after fuel cell failure	140
84	May 15, 1997	STS-84	*Atlantis*	9 days, 5 hours	Kennedy	Shuttle-Mir docking	141
85	July 1, 1997	STS-94	*Columbia*	15 days, 16 hours	Kennedy	Spacelab mission	142
86	Aug. 7, 1997	STS-85	*Discovery*	11 days, 20 hours	Kennedy	CRISTA-SPAS astronomy instruments	142
87	Sept. 25, 1997	STS-86	*Atlantis*	10 days, 19 hours	Kennedy	Shuttle-Mir docking	143
88	Nov. 19, 1997	STS-87	*Columbia*	15 days, 16 hours	Kennedy	Microgravity experiments; SPARTAN solar physics	144
89	Jan. 22, 1998	STS-89	*Endeavour*	8 days, 19 hours	Kennedy	Shuttle-Mir docking	146
90	April 17, 1998	STS-90	*Columbia*	15 days, 21 hours	Kennedy	Spacelab mission	147
91	June 2, 1998	STS-91	*Discovery*	9 days, 19 hours	Kennedy	Final Shuttle-Mir docking	148
92	Oct. 29, 1998	STS-95	*Discovery*	8 days, 21 hours	Kennedy	Former Mercury astronaut John Glenn aboard	150
93	Dec. 4, 1998	STS-88	*Endeavour*	11 days, 19 hours	Kennedy	NASA's First ISS assembly flight, carrying Node 1	160
94	May 27, 1999	STS-96	*Discovery*	9 days, 19 hours	Kennedy	ISS assembly & supply mission, using SpaceHab	164
95	July 23, 1999	STS-93	*Columbia*	4 days, 22 hours	Kennedy	Chandra X-ray Observatory launch	166
96	Dec. 19, 1999	STS-103	*Discovery*	7 days, 23 hours	Kennedy	Hubble Space Telescope servicing mission	168
97	Feb. 11, 2000	STS-99	*Endeavour*	11 days, 5 hours	Kennedy	Radar Topography earth observation	172
98	May 19, 2000	STS-101	*Atlantis*	9 days, 21 hours	Kennedy	ISS supply mission	173
99	Sept. 8, 2000	STS-106	*Atlantis*	11 days, 19 hours	Kennedy	ISS supply mission	174
100	Oct. 11, 2000	STS-92	*Discovery*	12 days, 21 hours	Edwards	ISS assembly flight, with Z1 truss	176
101	Nov. 30, 2000	STS-97	*Endeavour*	10 days, 19 hours	Kennedy	ISS assembly flight, with solar arrays and radiators	178
102	Feb. 7, 2001	STS-98	*Atlantis*	12 days, 21 hours	Edwards	ISS assembly flight, carrying the Destiny laboratory	179
103	March 8, 2001	STS-102	*Discovery*	12 days, 19 hours	Kennedy	ISS supply and crew rotation	181
104	April 19, 2001	STS-100	*Endeavour*	11 days, 21 hours	Edwards	ISS assembly flight, including a robotic arm for ISS	184
105	July 12, 2001	STS-104	*Atlantis*	12 days, 18 hours	Kennedy	ISS assembly flight, with the Quest airlock	186
106	Aug. 10, 2001	STS-105	*Discovery*	11 days, 21 hours	Kennedy	ISS supply and crew rotation	188
107	Dec. 5, 2001	STS-108	*Endeavour*	11 days, 19 hours	Kennedy	ISS supply and crew rotation	194

Sequence	Launch Date	STS number	Orbiter	Duration	Landing site	Mission highlights	Page
108	March 1, 2002	STS-109	Columbia	10 days, 22 hours	Kennedy	Hubble Space Telescope servicing mission	196
109	April 8, 2002	STS-110	Atlantis	10 days, 19 hours	Kennedy	ISS assembly flight, with S0 truss element	198
110	June 5, 2002	STS-111	Endeavour	13 days, 20 hours	Edwards	ISS supply; Mobile Base System; crew rotation	202
111	Oct. 7, 2002	STS-112	Atlantis	10 days, 19 hours	Kennedy	ISS assembly flight, with S1 truss element	204
112	Nov. 23, 2002	STS-113	Endeavour	13 days, 18 hours	Kennedy	ISS assembly flight; P1 truss; crew rotation	206
113	Jan. 16, 2003	STS-107	Columbia	15 days, 22 hours	n/a	Microgravity science; loss of vehicle during reentry	216
114	July 26, 2005	STS-114	Discovery	13 days, 21 hours	Edwards	ISS resupply, using Raffaello logistics module	218
115	July 4, 2006	STS-121	Discovery	12 days, 18 hours	Kennedy	ISS supply; crew rotation; Leonardo logistics module	222
116	Sept. 9, 2006	STS-115	Atlantis	11 days, 19 hours	Kennedy	ISS assembly flight; truss element; solar arrays	224
117	Dec. 9, 2006	STS-116	Discovery	12 days, 21 hours	Kennedy	ISS assembly; truss; SpaceHab; crew rotation	226
118	June 8, 2007	STS-117	Atlantis	13 days, 20 hours	Edwards	ISS assembly; truss; solar arrays; crew rotation	228
119	Aug. 8, 2007	STS-118	Endeavour	12 days, 18 hours	Kennedy	ISS assembly; truss element; SpaceHab	230
120	Oct. 23, 2007	STS-120	Discovery	15 days, 2 hours	Kennedy	ISS assembly flight; Harmony module; crew rotation	232
121	Feb. 7, 2008	STS-122	Atlantis	12 days, 18 hours	Kennedy	ISS assembly; Columbus laboratory; crew rotation	235
122	March 11, 2008	STS-123	Endeavour	15 days, 18 hours	Kennedy	ISS; Japanese Experiment Module; crew rotation	238
123	May 31, 2008	STS-124	Discovery	13 days, 18 hours	Kennedy	ISS; Japanese Kibo logistics module	240
124	Nov. 14, 2008	STS-126	Endeavour	15 days, 20 hours	Edwards	ISS servicing; solar array repairs; crew rotation	243
125	March 15, 2009	STS-119	Discovery	12 days,19 hours	Kennedy	ISS assembly flight; truss element; solar arrays	244
126	May 11, 2009	STS-125	Atlantis	12 days, 21 hours	Edwards	Final Hubble Space Telescope servicing mission	247
127	July 15, 2009	STS-127	Endeavour	15 days, 16 hours	Kennedy	ISS assembly, with the JEM Exposed Facility	250
128	Aug. 28, 2009	STS-128	Discovery	13 days, 21 hours	Edwards	ISS assembly flight, using Leonardo logistics module	255
129	Nov. 16, 2009	STS-129	Atlantis	10 days, 19 hours	Kennedy	ISS assembly, with ExPRESS logistics carriers	256
130	Feb. 8, 2010	STS-130	Endeavour	13 days, 18 hours	Kennedy	ISS assembly flight with Node 3 and Cupola	260
131	April 5, 2010	STS-131	Discovery	15 days, 3 hours	Kennedy	ISS supply, via Leonardo logistics module	263
132	May 14, 2010	STS-132	Atlantis	11 days, 18 hours	Kennedy	ISS assembly, with Russian Mini-Research Module	266
133	Feb. 24, 2011	STS-133	Discovery	12 days, 19 hours	Kennedy	ISS assembly; Leonardo permanently docked	272
134	May 16, 2011	STS-134	Endeavour	15 days, 17 hours	Kennedy	ISS assembly, with Alpha Magnetic Spectrometer	275
135	Summer 2011	STS-135	Atlantis	12 days	Kennedy	The final planned mission, dedicated to ISS supply	280

All images in this book are provided by NASA.
Very many thanks to Connie Moore and Margaret Persinger at NASA
for their patience and help in sourcing images.

Also by Piers Bizony from Zenith Press

ISBN: 978-0-7603-3710-3

CELEBRATING THE 40TH ANNIVERSARY OF THE FIRST MEN ON THE MOON

ONE GIANT LEAP

APOLLO 11 REMEMBERED

PIERS BIZONY

ML   11/11